Preface

This book contains all the work on Calculations, Science and Drawing that is needed for the Common Subject of the Mechanical Engineering and Mechanical Engineering Craft Studies Course, Part 3 (CGLI subject number 205). It is part 3 of the series of books which cater for the Craft Studies courses.

The work has been split up into fourteen chapters, four on Calculations, four on Science, three on Mechanisms, two on Materials and one on Drawing. With each chapter there is a generous number of graded exercises which is intended to help the student gain a better understanding of the topic.

Many of the worked examples in the text are essentially practical and they may be used for the Associated Studies part of the course.

I wish to thank the many colleagues and friends who have helped with their advice and criticism; the firms who have allowed me to use their diagrams; the publishers who have given me a considerable amount of help in the preparation of the manuscript.

Figure 5.2 is from the British Standards on screw threads; figs.5.4 to 5.6 and Table 11 are from BS 4500:1969; figs.5.7 to 5.25 are from BS 308:Part 3:1972; figs.5.32, 5.34 to 5.36 are from BS 499:1965. All are reproduced with the permission of the British Standards Institution, 2 Park Street, London W1A 2BS, from whom copies of the standards can be obtained. The cover has diagrams reproduced with the permission of Dean, Smith & Grace Ltd.

A.G.

Contents

Part 1 Calculations

1.1 Logarithms

The elementary concepts of logarithms were dealt with in *Mechanical Engineering Craft Studies Part 1*, (page 20). Before attempting more advanced work it will be as well to summarise the basic ideas.

1. A logarithm consists of two parts:
   (a) a whole number part called the *characteristic*,
   (b) a decimal part called the *mantissa* which is found directly from the log tables.

   The characteristic depends upon the size of the number. In the case of numbers greater than 1 it is found by subtracting 1 from the number of figures to the left of the decimal point in the given number. Thus

   log 2·53 = 0·4031
   log 25·3 = 1·4031
   log 253 = 2·4031

   In the case of numbers less than 1 (i.e. purely decimal numbers) the characteristic is always a negative whole number. Its value is numerically 1 more than the number of zeros following the decimal point. Thus

   log 0·253 = $\bar{1}$·4031
   log 0·0253 = $\bar{2}$·4031
   log 0·00253 = $\bar{3}$·4031

   Note that the minus sign is written above the characteristic. It must be clearly understood that

   $\bar{1}$·4031 = −1+0·4031
   $\bar{4}$·6258 = −4+0·6258

2. The table of anti-logarithms contains the numbers corresponding to the given logarithms. Only the mantissa is used when looking up anti-logs.

   In the case of a logarithm with a positive characteristic the number of figures to the left of the decimal point is found by adding one to the characteristic of the logarithm. Thus

   the number whose log is 2·4745 is 298·2
   the number whose log is 0·5082 is 3·222

   When the logarithm has a negative characteristic the number of zeros following the decimal point is 1 less than the numerical value of the negative characteristic. Thus

   the number whose log is $\bar{2}$·5326 is 0·03409
   the number whose log is $\bar{4}$·2618 is 0·000 182 7

3. To *multiply* find the logs of the numbers to be multiplied and *add* them together. The answer to the multiplication is obtained by finding the anti-log of the sum.

Example. Find the value of 12·86×0·0378

Number	log	
12·86	1·1093	
0·0378	$\bar{2}$·5775	ADD
Answer = 0·4862	anti-log $\bar{1}$·6868	

4. To *divide* find the log of each number and *subtract* the log of the denominator from the log of the numerator. The answer is obtained by finding the anti-log of the difference.

Example. Find the value of $\dfrac{72·35}{5·81}$

Number	log	
72·35	1·8594	
5·81	0·7642	SUBTRACT
Answer = 12·45	anti-log 1·0952	

Exercise 1.1

1. Use logarithms to find the values of the following:
   (a) 17·63×20·54
   (b) 328·4×54·7
   (c) 6819×1·285×17
   (d) 0·063 2×5·635
   (e) 0·006 418×0·072 43
   (f) 52·83×0·5219×081 92
   (g) 8·165÷3·142
   (h) 12·95÷128·3
   (i) $\dfrac{95·83×6·138}{8·179}$
   (j) $\dfrac{120·2}{9·125×123}$
   (k) $\dfrac{7·652×0·003 17}{2·561×0·06}$
   (l) $\dfrac{0·0169×0·002 875×0·9613}{7·325×5·362×0·1782}$

Powers of numbers. The quantity 3×3×3×3 may be written 3^4 which is called the fourth power of 3. The smaller number 4, which gives the number of threes to be multiplied together is called the index (plural: indices).

To find the value of a power of a number take the logarithm of the number and multiply it by the number which denotes the power. The value of the number raised to the given power is found by reading the anti-log of this product.

Example. Find the value of $(12·14)^2$
    log 12·14   = 1·0842
    log $(12·14)^2$ = 2×1·0842 = 2·1684
    By taking the anti-log of 2·1684:
        $(12·14)^2$ = 147·4

Example. Find the value of $(0·082 15)^5$
    log 0·082 15   = $\bar{2}$·9146
    log $(0·082 15)^5$ =5×$\bar{2}$·9146
                = 5×$\bar{2}$+(5×0·9146)
                = $\overline{10}$+4·5730
                = $\bar{6}$·5730
    By taking the anti-log of $\bar{6}$·5730
        $(0·082 15)^5$ = 0·000 003 741

Exercise 1.2
Use logarithms to find the values of the following:
1. $(29·38)^2$ 2. $(7·326)^3$
3. $(0·785)^3$ 4. $(0·036 14)^2$
5. $(0·1532)^4$

Roots of numbers. To find the value of a root of a number find the log of the number and divide it by the number denoting the root. The result obtained by this division is the log of the required root and its anti-log is the required root.

Example. Find the value of $\sqrt[3]{12·58}$
    log 12·58   = 1·0997
    log $\sqrt[3]{12·58}$ = 1·0997÷3 = 0·3666
    By taking the anti-log of 0·3666
        $\sqrt[3]{12·58}$ = 2·326

Example. Find the value of $\sqrt[3]{0·4806}$
    log 0·4806   = $\bar{1}$·6818

    log $\sqrt[3]{0·4806}$ = $\dfrac{\bar{1}·6818}{3}$ = $\dfrac{\bar{1}+0·6818}{3}$

We must make the negative characteristic exactly divisible by 3 and so we write

$$\frac{\bar{1}·6818}{3} = \frac{\bar{3}+2·6818}{3} = \bar{1}+0·8940 = \bar{1}·8940$$

By taking the anti-log of $\bar{1}$·8940
        $\sqrt[3]{0·4806}$ = 0·7833

Exercise 1.3
Find the values for the following:
1. $\sqrt[3]{15·38}$ 2. $\sqrt[4]{1·295}$
3. $\sqrt{0·2569}$ 4. $\sqrt[3]{0·06987}$
5. $\sqrt[3]{0·000 781 6}$

Examples involving powers and roots.
1. Find the value of $(14·68)^2 × (1·872)^3$
    In all calculations involving logarithms it pays to set out the work in tabular form. By doing this any errors that are made can be easily found and eliminated.

Number	operation	log	
$(14·68)^2$	2×1·1667	2·3334	ADD
$(1·872)^3$	3×0·2723	0·8169	
Answer = 1415	anti-log ←	3·1506	

The worst mistake that can be made in a calculation is that of misplacing the decimal point. To prevent this happening it is always worth while doing a rough check by using approximate numbers. When doing a rough check always try to select numbers so that they multiply out easily or so that they cancel.
For a rough estimate we will take
        $14^2 × 2^3$ = 196×8 = 1568
Although there is a considerable difference between the approximate answer of 1568 and the correct answer of 1415 the rough check shows the answer is 1415 and not 141·5 or 1415·0.

2. Find the value of $\sqrt{\dfrac{(0·1783)^3×212·5}{(0·057 21)^2×92·34}}$

Number	operation	log	Number	operation	lo
$(0·1783)^3$	3×$\bar{1}$·2512	$\bar{3}$·7536	$(0·057 21)^2$	2×$\bar{2}$·7575	$\bar{3}$·5
212·5		2·3274	92·34		1·9
Numerator		0·0810	Denominator		$\bar{1}$·4
Denominator		$\bar{1}$·4804 ←			
Answer =1·997	anti-log ←	0·6006 ÷ 2 = 0·3003			

$$\text{Rough check} = \sqrt{\frac{(0·2)^3×200}{(0·06)^2×100}} = \sqrt{\frac{0·008×200}{0·0036×100}}$$

$$= \sqrt{\frac{0·008×200}{0·004×100}} = \sqrt{4} = 2$$

since the rough check and the answer compare favourably there is no doubt that the decimal point is in the correct position in the answer.

Exercise 1.4
Find the values for the following:
1. $(2·998)^2×11·35$ 2. $(0·6257)^3×27·89$
3. $(16·29)^3÷89·76$ 4. $73·25÷(3·924)^3$
5. $\dfrac{3·215×(0·032)^4}{0·7918}$ 6. $\dfrac{50·46×(7·15)^3}{(0·7321)^2}$

7. $\sqrt[3]{0·1986}×(5·39)^2$ 8. $\sqrt{\dfrac{7·635}{2·152}}$

9. $\sqrt[3]{\dfrac{0·6372×5·618}{8·931×2·635}}$ 10. $\dfrac{6·918×0·2170}{\sqrt[3]{0·6183}}$

Fractional indices. These sometimes occur in engineering formulae and the method used is shown in the following examples

Example. Evaluate $18·27^{2/3}$
    log 18·27   = 1·2617

    log $18·27^{2/3}$ = $\dfrac{2}{3}$×1·2617 = $\dfrac{2×1·2617}{3}$

            = $\dfrac{2·5234}{3}$ = 0·8412

By taking the anti-log of 0·8412
        $18·27^{2/3}$ = 6·937

Example. Evaluate $0.0682^{3/4}$

$$\log 0.0682 \quad = \overline{2}.8338$$

$$\log 0.0682^{3/4} = \frac{3}{4} \times \overline{2}.8338 = \frac{3 \times \overline{2}.8338}{4}$$

$$= \frac{3 \times \overline{2} + 3 \times 0.8338}{4}$$

$$= \frac{\overline{6} + 2.5014}{4} = \frac{\overline{4}.5014}{4}$$

$$= \overline{1}.1253$$

By taking the anti-log of $\overline{1}.1253$

$$0.0682^{3/4} = 0.1335$$

Decimal indices. One way of dealing with these is to make the decimal index into a fractional index as shown in the following examples.

Example. Evaluate $7.016^{0.3}$

$$\text{since } 0.3 = \frac{3}{10}$$

$$7.016^{0.3} = 7.016^{3/10}$$

$$\log 7.016 \quad = 0.8461$$

$$\log 7.016^{3/10} = \frac{3 \times 0.8461}{10} = \frac{2.5383}{10}$$

$$= 0.2538 \text{ (to four significant figures)}$$

Taking the anti-log of 0.2538

$$7.016^{0.3} = 1.794$$

Example. Evaluate $0.3517^{0.12}$

$$\text{since } 0.12 = \frac{12}{100}$$

$$0.3517^{0.12} = 0.3517^{12/100}$$

$$\log 0.3517 \quad = \overline{1}.5462$$

$$\log 0.3517^{12/100} = \frac{12 \times \overline{1}.5462}{100}$$

$$= \frac{(12 \times \overline{1}) + (12 \times 0.5462)}{100}$$

$$= \frac{\overline{12} + 6.5544}{100}$$

$$= \frac{\overline{6} + 0.5544}{100}$$

$$= \frac{\overline{100} + 94.5544}{100}$$

$$= \overline{1} + 0.9455 = \overline{1}.9455$$

$$\therefore 0.3517^{0.12} = 0.8820$$

Exercise 1.5
Find values for the following:

1. $21.76^{1/3}$　　　　　2. $926.3^{2/5}$
3. $(0.7198)^{1/4}$　　　4. $(17.36)^{3/4}$
5. $(0.018\,73)^{2/3}$　　6. $(11.81)^{0.2}$
7. $(0.1673)^{0.4}$　　　8. $(0.0827)^{0.26}$
9. $(125.7)^{0.1}$　　　　10. $(0.0052)^{0.8}$

Evaluating formulae. Formulae are evaluated by substituting numerical values for the symbols as shown in the following examples.

Example. The out of balance force (in newtons) on a grinding wheel is given by the formula

$$F = \frac{m\pi^2 N^2 r}{900}$$

where m = mass of grinding wheel (kilograms)
$\quad\quad N$ = speed of wheel (revolutions per minute)
$\quad\quad r$ = distance of centre of gravity of wheel from the axis of rotation (metres)

Find the out of balance force when $m = 8.85$ kg, $N = 5000$ rev/min and $r = 0.1$ mm.

Substituting the given values and noting that $r = 0.0001$ m

$$F = \frac{8.85 \times (3.142)^2 \times 5000^2 \times 0.0001}{900}$$

Number	operation	log
8.85		0.9469
$(3.142)^2$	2×0.4972	0.9944
5000^2	2×3.6990	7.3980
0.0001		$\overline{4}.0000$
Numerator		5.3393
900		2.9542
Answer = 243.8	← anti-log	2.3851

$$\text{Rough check} = \frac{10 \times 3^2 \times 5000^2 \times 0.0001}{900} = 250$$

The rough check and the answer compare favourably and hence the out of balance force is $\underline{243.8 \text{ N}}$.

Example. The radius of a sphere, r centimetres, whose volume is V cubic centimetres is given by the formula

$$r = \sqrt[3]{\frac{3V}{4\pi}}$$

Find the value of r when $V = 98.16$ cm³. Substituting the given value of V in the formula

$$r = \sqrt[3]{\frac{3 \times 98.16}{4 \times 3.142}}$$

Number	log	Number	log
3	0.4771	4	0.6021
98.16	1.9920	3.142	0.4972
Numerator	2.4691	Denominator	1.0993
Denominator	1.0993		
	$1.3698 \div 3 = 0.4433$		

By taking the anti-log of 0.4433 the answer is 2.775 cm.

$$\text{Rough check} = \sqrt[3]{\frac{3 \times 100}{4 \times 3}} = \sqrt[3]{25} = 3 \text{ (approx.)}$$

Hence the value of r is $\underline{2.775 \text{ cm}}$

Example. The torque (newton metres) required to operate a drill may be obtained from the formula

$$T = Cf^{0.75}d^{1.8}$$

where C = a constant whose value depends on the metal being cut,

   f = the feed of the drill (millimetres per revolution)

and d = the drill diameter (millimetres).

Calculate the torque required to drill a 20 mm diameter hole in cast iron with a feed of 0·25 mm/rev. (The value of C when drilling cast iron is 0·07.)

Substituting the given values in the formula

$$T = 0.07 \times 0.25^{0.75} \times 20^{1.8}$$

Number	operation	log
0·07		$\bar{2}$·8451
0·25⁰·⁷⁵	$0.75 \times \bar{1}.3979$	$\bar{1}$·5484
20¹·⁸	1.8×1.3010	2·3418
Answer = 5·437	← anti-log	0·7353

Hence the torque required is 5·437 Nm

Exercise 1.6

1. The power (kilowatts) required when cutting cast iron is given by the formula

$$P = \frac{SDf}{18}$$

   where S = the cutting speed (metres per minute)

      D = the depth of cut (millimetres)

   and f = the feed of the tool (millimetres per revolution).

   Calculate the power required when cutting cast iron at a speed of 25 m/min with a feed of 0·5 mm/rev and a depth of cut of 4·5 mm.

2. The mean diameter of a taper pin is given by the formula $d = 1.13 \times \sqrt{\dfrac{PR}{DS}}$. Find the value of d when $P = 263$, $D = 63.51$, $R = 914.4$ and $S = 34\ 470$.

3. The volume of metal removed (cubic centimetres per minute) when drilling is given by the formula $V = \pi r^2 f / 1000$ where r = the radius of the hole (millimetres) and f = the feed of the drill (millimetres per minute). Find the volume of metal removed when drilling a hole 15 mm diameter with a feed of 250 mm/min.

4. When cutting mild steel a suitable cutting speed, v metres per minute, may be obtained by using the formula $v = C/T^{1/8}$

   where C = a constant

   and T = the tool life (minutes).

   Calculate a suitable cutting speed to give a tool life of 200 min if the value of C is 40.

5. The power (kilowatts) required when drilling a hole is given by the formula $P = 2\pi NT/60\ 000$,

   where N = the drill speed (revolutions per minute)

   and T = the torque on the drill (newton metres).

   Find the power required when $N = 2000$ rev/min and $T = 32$ Nm.

6. The coefficient of friction in a bearing with film lubrication is given by $\mu = K\sqrt{v/p}$

   where K = a constant

      v = the surface speed of the shaft (metres per second)

   and p = the bearing pressure (newtons per square millimetre).

   Find the value of μ when $K = 0.03$, $v = 1.5$ m/s and $p = 2.5$ N/mm².

7. The formula $M = 1.73 \times {}^3\sqrt{d^2/t}$ is used in connection with short tubes. Find M when $d = 2.8$ and $t = 0.56$.

8. The torque (newton metres) required to operate a drill is approximately given by $T = Cf^{3/4}d^2$. Find the value of T when $C = 0.11$, $f = 0.5$ and $d = 30$.

9. For a certain machining operation the relationship between the cutting speed, v metres per minute, and the tool life, T minutes, is given by $v = 35/T^{1/6}$. Find a suitable cutting speed to give a tool life of 180 min.

10. When the diameter, D, of a standard c shaft is less than 40 mm the fundamental deviation (micrometres) is calculated from $f = 52D^{0.2}$. Find the value of f when $D = 25$ mm.

11. For a standard M hole the fundamental deviation is calculated from $f = 2.8 \times \sqrt[3]{D}$ where D is the hole diameter (millimetres) and f is in micrometres. Find f when $D = 39.21$ mm.

Accuracy of the log tables. When using four-figure log tables it should be realised that the fourth significant figure is suspect. As a general rule four-figure tables should only be used when an accuracy of three significant figures in the answer is satisfactory. This degree of accuracy is not acceptable in many cases, particularly when dealing with precision measurement.

As an example: $11.43 \times 0.4628 = 52.90$ to four significant figures by ordinary multiplication and 52·88 by using four-figure log tables.

Five-figure log tables are more accurate than the four-figure type but in five-figure tables the fifth significant figure is suspect. For the calculation above the five-figure tables give the answer as 52·894 whilst the correct answer is 52·898 correct to five significant figures.

If very accurate calculations are required then either seven-figure log tables or a calculating machine should be used.

Formulae containing reciprocals, squares and square roots. When formulae contain reciprocals, squares or square roots use the appropriate tables which are usually found in books of logarithmic and other tables.

Example. Find the value of

$$\frac{1}{8{\cdot}916}+\frac{1}{(7{\cdot}23)^2}+\frac{1}{\sqrt{7{\cdot}46}}$$

using the table of squares: $(7{\cdot}23)^2 = 52{\cdot}27$
using the table of square roots: $\sqrt{7{\cdot}46} = 2{\cdot}731$
The calculation then becomes

$$\frac{1}{8{\cdot}916}+\frac{1}{52{\cdot}27}+\frac{1}{2{\cdot}731}$$

and using the table of reciprocals we obtain

$$\frac{1}{8{\cdot}916}+\frac{1}{52{\cdot}27}+\frac{1}{2{\cdot}731} = 0{\cdot}1121+0{\cdot}0191+0{\cdot}3662$$

$$= \underline{0{\cdot}4974}$$

Example. The Brinell hardness number of a material is given by the formula

$$\frac{2P}{\pi D[D- \sqrt{D^2-d^2}]}$$

Calculate the Brinell hardness number when $P = 3000$, $D = 10$ and $d = 4{\cdot}26$. Take $\pi = 3{\cdot}142$.
Substituting the given figures into the formula we have
Brinell hardness number

$$= \frac{2\times3000}{3{\cdot}142\times10\times[10-\sqrt{10^2-4{\cdot}26^2}]}$$

$$= \frac{6000}{31{\cdot}42[10-\sqrt{100-18{\cdot}15}]}$$

$$= \frac{6000}{31{\cdot}42\times[10-\sqrt{81{\cdot}85}]}$$

$$= \frac{6000}{31{\cdot}42\times[10-9{\cdot}047]}$$

$$= \frac{6000}{31{\cdot}42\times0{\cdot}953}$$

It is now possible to use the log tables because we have only multiplication and division signs left in the calculation.

Number	log	Number	log
6000	3·7782	31·42	1·4972
Denom.	1·4763	0·953	$\bar{1}$·9791
200·4	2·3019	Denom. 1·4763	

Thus the Brinell hardness number is 200·4

Exercise 1.7

1. When checking the radius of large round work the formula $R = (M-d)^2/8d$ is used. Find R when $M = 79{\cdot}25$ and $d = 12$.

2. When actual measurement is impossible the radius of a circular groove may be found by using the formula $R = (H/2) + (L^2/8H)$. Find R when $H = 26{\cdot}17$ and $L = 88{\cdot}90$.

3. If $C = 2\times \sqrt{2hr-h^2}$ find the value of C when $h = 7$ and $r = 5$.

4. If $s = \pi(4h^2+c^2)/4$ find s when $h = 19{\cdot}28$ and $c = 5{\cdot}71$.

5. Using, where necessary, tables of squares, square roots, reciprocals and logarithms evaluate the following:

   (a) $\sqrt{(17{\cdot}6)^2+(11{\cdot}7)^2}$

   (b) $\sqrt{(0{\cdot}016\,89)^2+(0{\cdot}2314)^2}$

   (c) $\dfrac{1}{8{\cdot}2}+\dfrac{1}{9{\cdot}9}$

   (d) $\dfrac{1}{0{\cdot}7325}-\dfrac{1}{0{\cdot}9817}$

   (e) $\dfrac{1}{\sqrt{7{\cdot}517}}+\dfrac{1}{(0{\cdot}829)^2}+\dfrac{1}{0{\cdot}0749}$

   (f) $\dfrac{1}{71{\cdot}36}+\dfrac{1}{\sqrt{863{\cdot}5}}+\dfrac{1}{(7{\cdot}589)^2}$

   (g) $\dfrac{\sqrt{18{\cdot}16}}{(5{\cdot}26)^2-(3{\cdot}17)^2}$

   (h) $\dfrac{\sqrt{(1{\cdot}82)^2-(0{\cdot}763)^2}}{2\times3{\cdot}18\times7{\cdot}25}$

1.2 Transposition of formulae

In the formula $C = nE/(R+nt)$, C is called the *subject* of the formula. It may be that we are given values of C, n, E and r and have to find the value of R. We can do this by substituting the given values in the formula and solving the resulting equation for R. The method is shown in the following example.

Example. Find R from the formula $C = nE/R+nr$ when $C = 1\cdot5$, $n = 7$, $E = 3$ and $r = 1\cdot1$. Substituting the given values in the formula we have:

$$1\cdot5 = \frac{7\times3}{R+7\times1\cdot1}$$
$$1\cdot5 = \frac{21}{R+7\cdot7}$$
$$1\cdot5(R+7\cdot7) = 21$$
$$1\cdot5R+11\cdot55 = 21$$
$$1\cdot5R = 21-11\cdot55$$
$$1\cdot5R = 9\cdot45$$
$$R = \frac{9\cdot45}{1\cdot5}$$
$$R = 6\cdot3$$

Transposition of formulae. Frequently several corresponding values of c, n, E and r are known and we want to find the corresponding values of R. We can use the method of the previous example but considerable time and effort will be spent in solving the resulting equations. Much of this time would be saved if we could express the formula with R as the subject because then we need only substitute the values of c, n, E and r in the rearranged formula.

The process of rearranging a formula, so that one of the other symbols becomes the subject, is called *transposing the formula*. The rules used in transposition are the same as those used in solving equations. The methods used are as follows:

Symbols connected as a product.
1. Transpose the formula $F = ma$ to make a the subject. Divide both sides by m, then
$$\frac{F}{m} = \frac{ma}{m}$$
Cancelling the m's on the right hand side
$$\frac{F}{m} = a \text{ or } a = \frac{F}{m}$$

2. Make h the subject of the formula $V = \pi r^2 h$. Divide both sides by πr^2, then
$$\frac{V}{\pi r^2} = \frac{\pi r^2 h}{\pi r^2}$$
Cancelling the πr^2's on the right hand side
$$\frac{V}{\pi r^2} = h \text{ or } h = \frac{V}{\pi r^2}$$
Symbols connected as a quotient.
1. Transpose $f = P/A$ to make P the subject. Multiply both sides by A, then
$$f\times A = \frac{P}{A}\times A$$
Cancelling the A's on the right hand side
$$fA = P \text{ or } P = fA$$
2. Transpose $H = C^2RT/J$ for J. Multiply both sides by J
$$H\times J = \frac{C^2RT}{J}\times J$$
Cancelling the J's on the right hand side
$$HJ = C^2RT$$
Dividing both sides by H
$$\frac{HJ}{H} = \frac{C^2RT}{H}$$
Cancelling the H's on the left hand side
$$J = \frac{C^2RT}{H}$$
Symbols connected by plus and minus signs.
Remember that when a term is transferred from one side of a formula to the other its sign is changed.
1. Transpose $T = t+273$ for t. Subtract 273 from both sides, then
$$T-273 = t \text{ or } t = T-273$$
2. Transpose $w = H+Cr$ for C. Subtract H from both sides, then
$$w-H = Cr$$
Divide both sides by r, then
$$\frac{w-H}{r} = \frac{Cr}{r}$$
Cancelling the r's on the right hand side
$$\frac{w-H}{r} = C \text{ or } C = \frac{w-H}{r}$$

Formulae containing brackets.

1. Transpose $D = d + (t/6)$ for t. Subtract d from both sides, then

$$D - d = \frac{t}{6}$$

Multiply both sides by 6, then

$$6 \times (D - d) = \frac{t}{6} \times 6$$

Cancelling the 6's on the right hand side

$$6(D - d) = t \text{ or } t = 6(D - d)$$

2. Transpose $l = a + (n-1)d$ for d. Subtract a from both sides, then

$$l - a = (n-1)d$$

Divide both sides by $(n-1)$, then

$$\frac{l-a}{(n-1)} = \frac{(n-1)d}{(n-1)}$$

Cancelling the $(n-1)$'s on the right hand side

$$\frac{l-a}{n-1} = d \text{ or } d = \frac{l-a}{n-1}$$

3. Transpose $Q = w(H-h)/(T-t)$ for H. Multiply both sides by $(T-t)$, then

$$Q \times (T-t) = \frac{w(H-h)}{(T-t)} \times (T-t)$$

Cancelling the $(T-t)$'s on the right hand side

$$Q(T-t) = w(H-h)$$

Dividing both sides by w

$$\frac{Q(T-t)}{w} = \frac{w(H-h)}{w}$$

Cancelling the w's on the right hand side

$$\frac{Q(T-t)}{w} = H-h$$

Adding h to both sides of the equation

$$\frac{Q(T-t)}{w} + h = H \text{ or } H = \frac{Q(T-t)}{w} + h$$

4. Transpose $x = n/(n-1)$ for n. Multiply both sides by $(n-1)$

$$x \times (n-1) = \frac{n}{(n-1)} \times (n-1)$$

Cancelling the $(n-1)$'s on the right hand side

$$x(n-1) = n$$

Remove bracket on left hand side

$$xn - x = n$$

Grouping the terms containing n on the left hand side

$$xn - n = x$$

Factorising the left hand side

$$n(x-1) = x$$

Dividing both sides by $(x-1)$

$$\frac{n(x-1)}{(x-1)} = \frac{x}{(x-1)}$$

$$n = \frac{x}{x-1}$$

Formulae containing roots and powers.

1. Transpose $H = C/N^2$ for N. Multiply both sides by N^2

$$HN^2 = C$$

Divide both sides by H

$$N^2 = \frac{C}{H}$$

Take the square root of both sides

$$N = \sqrt{\frac{C}{H}}$$

2. Transpose $d = \sqrt{2hr}$ for r. Square both sides of the equation

$$d^2 = 2hr$$

Divide both sides by $2h$

$$\frac{d^2}{2h} = r \text{ or } r = \frac{d^2}{2h}$$

Exercise 1.8

Transpose the following formulae making the symbol stated the subject.

1. $C = \pi d$ for d
2. $S = \pi dn$ for n
3. $PV = c$ for V
4. $A = \pi r l$ for r
5. $v^2 = 2gh$ for h
6. $I = PRT$ for R
7. $P = \dfrac{I}{N}$ for N
8. $I = \dfrac{E}{R}$ for R
9. $D = \dfrac{M}{V}$ for M
10. $P = \dfrac{RT}{V}$ for T
11. $d = \dfrac{0.866}{N}$ for N
12. $S = \dfrac{ts}{T}$ for T
13. $V = \dfrac{\pi d^2 h}{4}$ for h

14. $h = \dfrac{V}{\pi r^2}$ for V

15. $v = u + at$ for t

16. $n = p + cr$ for r

17. $y = ax + b$ for x

18. $H = S + qL$ for q

19. $V = \dfrac{2R}{R - r}$ for r

20. $C = \dfrac{E}{R + r}$ for E

21. $S = \pi r(r + h)$ for h

22. $H = ws(T - t)$ for T

23. $C = \dfrac{N - n}{2p}$ for N

24. $T = \dfrac{12(D - d)}{L}$ for d

25. $V = \dfrac{2R}{R - r}$ for r

26. $P = \dfrac{S(C - F)}{C}$ for F

27. $v = \sqrt{2gh}$ for h

28. $E = \frac{1}{2}mv^2$ for v

29. $A = \pi r^2$ for r

30. $w = 1000\sqrt{d}$ for d

31. $t = 2\pi\sqrt{\dfrac{l}{g}}$ for l

32. $t = 2\pi\sqrt{\dfrac{W}{gf}}$ for f

33. $a^2 = b^2 + c^2$ for c

34. $D = 1 \cdot 2\sqrt{dL}$ for L

35. In the formula $v^2 = u^2 + 2as$ find s when $v = 30$, $u = 20$ and $a = 2$.

36. If two resistances are connected in parallel the total resistance R ohms is given by the formula:
$$\frac{1}{R} = \frac{1}{R_1} + \frac{1}{R_2}$$
Find R when $R_1 = 3$ ohms and $R_2 = 5$ ohms.

37. In the formula $W = \dfrac{CV}{1000}$ find the value of V when $C = 15$ and $W = 3 \cdot 6$.

38. If $H = h + \dfrac{v^2}{2g}$ find v when $H = 8$, $h = 3$ and $g = 10$.

39. When measuring the diameter of a large bore the formula $D = L + \dfrac{w^2}{2L}$ is used. Find the value of w when $D = 381 \cdot 2$ mm and $L = 380$ mm.

40. In measuring the diameter of a large cylinder the formula $R = \dfrac{(M - d)^2}{8d}$ is used. Find the value of M when $d = 12 \cdot 00$ mm and $R = 88 \cdot 00$ mm.

41. Find the value of l from the formula $k = \sqrt{\dfrac{l}{m}}$ when $k = 375$ and $m = 24$.

42. Find the value of s from the formula
$$p = \frac{f}{2} + \frac{f}{2} \times \sqrt{1 + \frac{4s^2}{f^2}}$$
when $f = 3050$ and $p = 4030$.

1.3 Charts and graphs

Graphs from tabular information. Data in handbooks and figures obtained as a result of an experiment are often stated in tabular form. The information is often made more understandable if it is made into a chart or a graph. When the results of an experiment are plotted the points often deviate slightly from a straight line or a smooth curve. We must expect this because of errors in measurement and observation. If the points when plotted show a trend towards a straight line or a smooth curve the best straight line or curve is drawn. Although the line will not pass through some of the points an attempt must be made to ensure an even spread of the points above and below the straight line or the curve.

Examples

1. During a test to find how the power of a lathe (P kilowatts) varied with the depth of cut (d millimetres) the results in the table below were obtained. The speed and feed of the lathe were kept constant.

d (mm)	0·5	1·0	1·5	2·0	2·5	3·0
P (kW)	0·94	1·06	1·18	1·31	1·46	1·59

Draw a graph of this information plotting P on the vertical axis and find the power required for a depth of cut of 2·15 mm.

The graph is shown plotted in Fig. 1.1 where it can be seen that the trend is towards a straight line. From the graph it is found that when the depth of cut is 2·15 mm the power required is 1·36 kW.

2. In order to establish the best rake angle for a lathe roughing tool, tests were made to find the power (P kilowatts) used when cutting with tools having various rake angles (θ degrees). The cutting speed, feed and depth of cut were all kept constant. The following readings were obtained:

$\theta°$	0	10	20	30	40	50
P (kW)	1·57	1·04	0·79	0·70	0·68	0·67

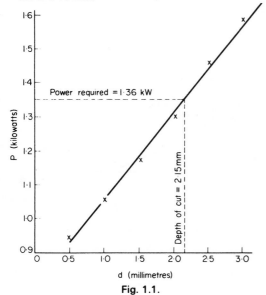

Fig. 1.1.

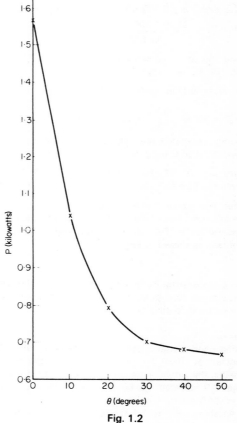

Fig. 1.2

Draw a graph of rake angle against power used, plotting the power vertically. From the graph suggest a practical rake angle.

The graph is plotted in Fig. 1.2 where it can be seen that the trend is towards a smooth curve. The graph shows that the power falls quickly at first as the rake angle increases and then tends to level off. Although the largest rake angle requires the least power a large rake angle always creates a weak tool. Hence a practical rake angle would be about 30°.

Graphs from formulae. When a formula is used a large amount of repetitive calculation can be avoided by drawing a graph.

Example. A formula used in calculating the spindle speed of a lathe is $N = \dfrac{1000\,S}{\pi D}$ where N is the spindle speed (revolutions per minute), S is the cutting speed (metres per minute) and D is the workpiece diameter (millimetres). Plot a graph to show the spindle speeds required when cutting cast iron at a cutting speed of 27 m/min. The graph should cover a range of workpiece diameters from 10 mm to 100 mm.

Substituting $\pi = 3 \cdot 142$ and $S = 27$ in the formula

$$N = \frac{1000 \times 27}{3 \cdot 142 \times D} = \frac{8594}{D}$$

We now decide on some values of D, say 10, 20, 30, 40, 50, 60, 70, 80, 90 and 100 mm and substitute each of these values of D in the formula $N = \dfrac{8594}{D}$. For instance,

when $D = 20$, $N = \dfrac{8594}{20} = 429 \cdot 7$.

Having calculated the values of N for each of the chosen values of D we draw up the following table:

D (mm)	10	20	30	40	50	60	70	80	90	100
N (rev/min)	859	430	287	215	172	143	123	108	96	86

When plotting graphs of formulae the single symbol on the left-hand side of the equation is, almost invariably, plotted vertically. Thus N is plotted vertically and D horizontally as shown in Fig. 1.3.

Break even points. By plotting two graphs on the same axes the break even point can be found. The method is shown in the following examples.

Examples.

1. Using hand methods the cost of making an article is constant at £0·80. By using special tools the cost of making the article falls uniformly from £1·50 each when 10 are made to £0·40 when 1000 are made. Find the break even point.

As shown in Fig. 1.4 the break even point occurs

when 640 articles are made. Hence unless more than 640 articles are to be made the special tools should not be used and production should be by hand methods only.

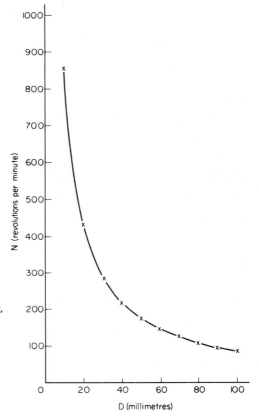

Fig. 1.3

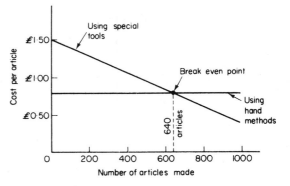

Fig. 1.4

2. A factory manufactures a product which is sold to retailers at £15 each. Production costs vary with numbers produced as shown in the table below.

Number produced per year	0	1000	2000	3000	4000	5000
Total cost in £ per year	15 000	33 000	46 000	54 000	61 000	65 000

Use this information to draw graphs on the same axes showing income and cost against the number produced per year. From the graphs find:

(a) the annual output for which income equals costs (i.e. the break even point)

(b) the output corresponding to the maximum loss per year

(c) the profit when 4500 units are produced per year. First draw up a table showing the number produced per year and total income per year.

Number produced per year	0	2000	5000
Total income in £ per year	0	30 000	75 000

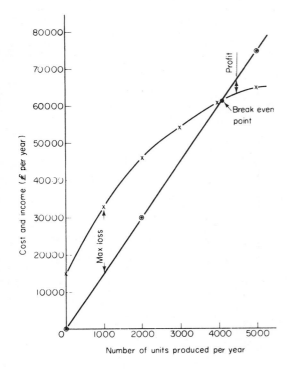

Fig. 1.5

The graphs are drawn in Fig. 1.5.

(a) The break even point is the intersection of the two graphs. It occurs when the number produced per year is 4100.

(b) If the number produced per year is less than 4100 then a loss is made. The maximum loss is the greatest distance between the two graphs. This occurs when the output is 1000 units per year.

(c) A profit occurs when the output is greater than 4100 units per year. The amount of the profit is the difference between the two lines. Thus the profit for an output of 4500 units per year is £4400.

Positive and negative values. It sometimes happens that in plotting a graph we have to include co-ordinates which are positive and negative. It is usual to call the horizontal axis the x-axis and the vertical axis the y-axis. The procedure shown in Fig. 1.6 is then adopted.

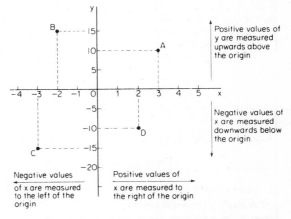

Fig. 1.6   Positive and negative co-ordinates

The co-ordinates for the four points A, B, C and D in Fig. 1.6 are as shown in the following table:

Point	x co-ordinate	y co-ordinate
A	3	10
B	−2	15
C	−3	−15
D	2	−10

Positive and negative values are often used to denote a change in direction. For instance a temperature of 50 °C above the freezing point of 0 °C would be +50 °C whilst a temperature of 20 °C below the freezing point would be −20 °C. A dimension of 0·03 mm above the nominal size would be +0·03 mm whilst a dimension of 0·05 mm below nominal would be −0·05 mm.

Example. A refrigerator converts a small amount of water at 25 °C into ice at −10 °C. It takes 5 min to reduce the temperature of the water to 0 °C and then for 15 min the temperature remains constant at 0 °C whilst the water is converted into ice. It then takes a further 2½ min for the ice to reach −10 °C. Show this information on a graph, plotting temperature vertically.

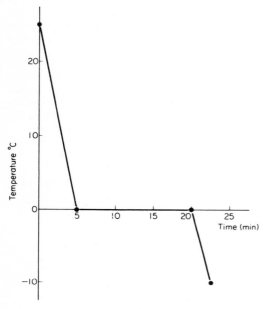

Fig. 1.7

The graph is shown in Fig. 1.7 where it is seen that the negative temperatures have been plotted below the horizontal axis.

Trends. The graphical representation of data will often detect a trend.

Example. Twenty successive turned pieces had their diameters measured with the following results which are stated in units of 0·01 mm above and below the nominal diameter of 20 mm.

Workpiece no.	1	2	3	4	5	6	7	8	9	10
diam. (0·1 mm)	−2	−1	−1	0	−1	0	+1	+1	+1	+2

Workpiece no.	11	12	13	14	15	16	17	18	19	20
diam. (0·1 mm)	+1	+2	+3	+3	+4	+5	+4	+4	+5	+6

Plot this information on a graph and state any conclusion that can be formed.

Fig. 1.8

The points are plotted in Fig. 1.8 and it is seen that they do not approximate either to a straight line or a smooth curve. Hence the best we can do is to join successive points by a straight line.

On looking at the diagram (Fig. 1.8) we notice that the trend is for the workpiece size to increase perhaps because of tool wear or some other cause. The main reason for drawing charts like this is to find out if there is a trend present or if there is anything abnormal about the readings.

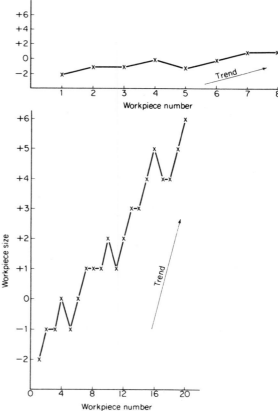

Fig. 1.9   The same chart with vastly differing scales

Effect of scales. The scales that are chosen for a chart or graph can accentuate the characteristics of the graph. For example, consider the chart of Fig. 1.8. If we shorten the horizontal scale and lengthen the vertical scale then the trend becomes greatly magnified. However, if we lengthen the horizontal scale and shorten the vertical scale the trend becomes hardly noticeable (see Fig. 1.9). Hence whenever looking at a chart look at the scales first before looking for any abnormalities.

Exercise 1.9

1. In order to achieve a tool life of 3 hours the cutting speeds for various depths of cut are as follows:

Depth of cut (mm)	3	4	6	8	12	16
Cutting speed (m/min)	160	140	117	104	88	79

Draw a graph with depth of cut horizontal and from it estimate the depth of cut for a cutting speed of 100 m/min. Estimate also the cutting speed when the depth of cut is 10 mm.

2. When cutting tool steel the following drill speeds are recommended:

Drill diameter (mm)	1	2	4	6	8	10	12	16
Drill speed (rev/min)	3056	1528	764	510	382	306	255	191

Draw a graph with the drill diameter horizontal and find the drill speed when using a 3 mm diameter drill and also a 9 mm diameter drill.

3. In an experiment to find the forces required to pull blocks of different weights along a horizontal surface the following results were obtained:

Weight of block (N)	4	8	12	16	20	24	28
Force required (N)	1·16	2·52	3·48	4·96	6·08	7·08	8·44

Plot these readings with weight of block horizontal and find the force required to pull a block weighing 17·6 N.

4. The following readings were obtained during a test to find the efficiency of lifting tackle:

Load (N)	100	200	300	400	500	600
Efficiency (%)	46	56·2	60·1	62·6	63·8	64

Plot a graph of load against efficiency (efficiency vertical) and estimate the efficiency when the load is 450 N.

5. The volume of a cylinder is found from the formula $V = \pi r^2 h$ where V is the volume (cubic millimetres), h is the height (millimetres) and r is the radius (millimetres). Draw a graph to show the relationship between volume and radius from 10 mm to 200 mm.

Take $\pi = \frac{22}{7}$ and the height = 14 mm.

6. The heat, H, produced by an electric current, I, is given by the formula $H = I^2 R$. If the resistance R is constant at 10 ohms draw a graph to show the heat produced for values of I between 0 and 5.

7. The Brinell hardness number of a material is given by the formula $\dfrac{2P}{\pi D [D - \sqrt{D^2 - d^2}]}$. If the load P is 3000 kg and the diameter of the ball D is 10 mm draw a graph showing the B.H.N. against the diameter of the impression d millimeters. Plot d horizontally and take its value as 3, 4, 5, 6 and 7 mm. Find the B.H.N. when the impression is 4·3 mm.

8. The approach distance of a milling cutter, A mm, the cutter diameter, C mm, and the depth of cut, D mm, are connected by the formula $A = \sqrt{C(D - C)}$. If the cutter diameter is 150 mm draw a graph of A against D taking values of D from 1 to 10 mm.

9. The cost of making an article by hand is constant at £1·20. By using special tools we can reduce the cost from £2·50 each when 20 are made to £0·80 when 200 are made. By drawing graphs find the break even point.

10. The cost of producing a batch of articles comprises a fixed charge of £150 plus 30p per article. If more advanced tooling is used the cost comprises a fixed charge of £400 plus 20p per article. By drawing graphs of number produced against total cost find the break even point. The graphs should cater for the number produced up to 4000.

11. A factory manufactures a product which is sold to retailers at £10 each. The cost of producing the articles comprises a fixed charge of £1500 plus a manufacturing cost of £5 per article. Use this information to draw graphs showing income and cost against the number produced. From the graphs find:
    (a) the output for which income equals costs
    (b) the loss when the output is 400
    (c) the profit when the output is 900.

12. Plot the points shown in the table below:

Point	x	y
A	−5	+3
B	2	−4
C	−3	−5
D	−2	−7
E	5	−2

13. Ice at −20 °C is heated for 5 min and its temperature rises uniformly to 0 °C. The temperature remains constant at 0 °C for 40 min whilst the ice turns into water and then rises uniformly for 50 min to 100 °C. The temperature then remains constant at 100 °C for 270 min whilst the water turns into steam. The steam is then superheated to a temperature of 200 °C

which takes a further 10 min. Show this information on a chart.

14. The table below gives the temperature of the air at 12.00 noon on seven successive days. Plot a graph to illustrate the information:

Day	June 6	7	8	9	10	11	12
Temp. (°C)	16	20	16	18	22	14	17

15. The lengths of 12 parts produced on a lathe were measured with the following results which are in units of 0·01 mm above and below 22·00 mm.

Workpiece no.	1	2	3	4	5	6
Workpiece length (0·01 mm)	+2	+1	+2	0	+1	0

Workpiece no.	7	8	9	10	11	12
Workpiece length (0·01 mm)	0	−1	0	−2	−3	−3

Plot this information on a chart and state if there is a trend present.

16. The diameters of 10 turned bars were measured with the following results. The measurements are stated in units of 0·01 mm above and below 35·00 mm.

Workpiece no.	1	2	3	4	5
Diameter (0·01 mm)	−0·4	−0·2	+1·4	+0·6	+1·2

Workpiece no.	6	7	8	9	10
Diameter (0·01 mm)	+1·6	−0·8	+0·2	−0·8	+0·8

Plot this information on a chart. Is any kind of a trend suggested by the chart?

Other kinds of charts. Many different kinds of charts are used to depict information.

Suppose that in a certain factory the number of persons employed in various jobs is as given in the table below:

Type of personnel	Number employed	Percentage of total employed
Machinists	140	35
Fitters	120	30
Clerical staff	80	20
Labourers	40	10
Draughtsmen	20	5
Total	400	100

The information in the table can be represented pictorially in several ways:

The pie chart. (Fig. 1.10) displays the proportions as angles (or sector areas) the complete circle representing the total number employed. Thus for machinists the angle of the sector is

$$\frac{140}{400} \times 360° = 126°$$

and for fitters

$$\frac{120}{400} \times 360° = 108° \text{ etc.}$$

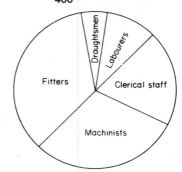

Fig. 1.10   The pie chart

The bar chart (Fig. 1.11) relies on heights (or areas) to

Draughtsmen	5 %
Labourers	10 %
Clerical staff	20 %
Fitters	30 %
Machinists	35 %

Fig. 1.11   The 100% bar chart

convey the proportions the total height of the diagram representing 100%.

The horizontal bar chart (Fig. 1.12) gives a better comparison of the various types of personnel employed but

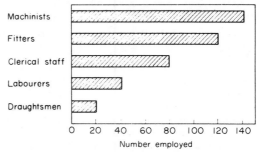

Fig. 1.12   Horizontal bar chart

it does not readily display the total number employed in the factory.

The histogram. This chart is used where a large amount of information has to be depicted. Suppose that the diameters of 100 ball bearings, nominally 15·00 mm diameter, are measured. We might obtain the following information:

Diameter (mm)	Number of ball bearings with this diameter
14·96	2
14·97	4
14·98	11
14·99	20
15·00	23
15·01	21
15·02	9
15·03	8
15·04	2

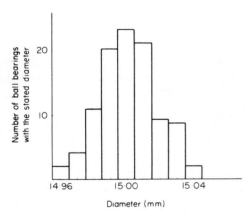

Fig. 1.13   The histogram

The table gives a good idea of how the sizes of the ball bearings vary but a histogram (Fig. 1.13) which consists of a set of rectangles, each of the same width and whose heights represent the numbers in the second column of the table, makes the information more understandable.

Exercise 1.10

1. A building contractor surveying his labour finds that 30% are engaged on factory work, 40% on house building and 30% on public works (schools, hospitals etc.). Depict this information on (a) a pie chart, (b) a single vertical bar chart.

2. A survey of the personnel in a certain factory gave the following information on the number of people employed: machinists 453, fitters 210, clerical staff 94, labourers 56, design staff 120 and production staff 75. Represent this information on (a) a pie chart (b) a horizontal bar chart.

3. An industrial organisation gives an aptitude test to all applicants for employment. The results for 150 people taking the test are as follows:

Score	1	2	3	4	5	6	7	8	9	10
Number of applicants with this score	6	12	15	21	35	24	20	10	6	1

Draw (a) a horizontal bar chart (b) a histogram.

4. The table below gives forging temperature ranges for various metals. Draw a horizontal bar chart to depict the information.

Metal	Forging range (°C)
High speed steel	850–1090
High carbon steel	770–1120
Medium carbon steel	770–1290
Wrought iron	860–1340
High tensile brass	600–800
Brass (60/40)	600–800
Copper	450–1000

5. A batch of 100 metal bars of nominal length 200 mm was measured to the nearest 0·1 mm with the results shown below:

Length (mm)	199·6	199·7	199·8	199·9	200·0
Number of bars with this length	2	4	12	18	31

Length (mm)	200·1	200·2	200·3	200·4
Number of bars with this length	22	8	2	1

Draw a histogram of this information.

6. During a mass production process 1000 samples, each of 40 items, are examined. The number of defectives in the samples are shown below.

Number of defectives in the sample	0	1	2	3	4
Number of samples with this number of defectives	620	282	80	16	2

Draw a histogram of this information.

1.4 Trigonometry

The definitions of the sine, cosine and tangent of an angle have been given in *Mechanical Engineering Craft Studies Part 2* (page 18). They are summarised below:
Using the notation of Fig. 1.14

$$\sin A = \frac{a}{b} \left(\text{i.e.} \ \frac{\text{opposite}}{\text{hypotenuse}} \right)$$

$$\cos A = \frac{c}{b} \left(\text{i.e.} \ \frac{\text{adjacent}}{\text{hypotenuse}} \right)$$

$$\tan A = \frac{a}{c} \left(\text{i.e.} \ \frac{\text{opposite}}{\text{adjacent}} \right)$$

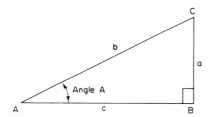

Fig. 1.14

Examples

1. Ten holes are equally spaced round a pitch circle of 200 mm diameter. Calculate the centre distance between adjacent holes.

    The pitch circle is shown in Fig. 1.15. Since AO = BO (radii), △AOB is isosceles.

$$\angle AOB = \frac{360°}{10} = 36°$$

Draw a perpendicular OC as shown

then $\angle AOC = \frac{1}{2} \angle AOB = 18°$

In △COA,

$$\frac{AC}{AO} = \sin \angle AOC$$

$$\frac{AC}{100} = \sin 18°$$

AC = 100 × sin 18° = 100 × 0·3090 = 30·90 mm

AB = 2 × AC = 2 × 30·90 = 61·80 mm

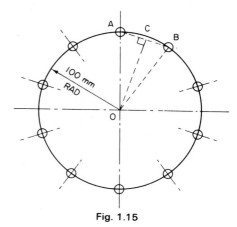

Fig. 1.15

2. Two holes are to be drilled in a plate as shown in Fig. 1.16. Calculate the dimension *a* that is needed for marking out purposes.

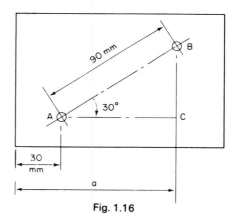

Fig. 1.16

In △ABC, $\angle ACB = 90°$

$\therefore \ \dfrac{AC}{AB} = \cos \angle BAC$

$$\frac{AC}{90} = \cos 60°$$

AC = 90 × cos 60° = 90 × 0·8660 = 77·94 mm

a = 30 + AC = 30 + 77·94 = 107·94 mm

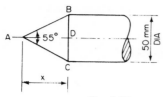

Fig. 1.17

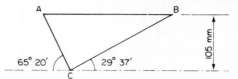

Fig. 1.20

3. The end of a bar is to be machined as shown in Fig. 1.17. Calculate the length x.

△ABC is isosceles. Hence the perpendicular AD bisects BC and ∠BAC.

In △ACD,

$$\angle CAD = 27° \ 30'$$

$$\angle ACD = 90° - 27° \ 30' = 62° \ 30'$$

$$CD = 25 \text{ mm}$$

$$\frac{AD}{CD} = \tan \angle ACD$$

$$\frac{AD}{25} = 62° \ 30'$$

$$AD = 25 \times 1·9210 = 48·025 \text{ mm}$$

$$\therefore \text{ length } x = 48·025 \text{ mm}$$

Exercise 1.11

1. A bar has to be machined as shown in Fig. 1.18. Calculate the angle φ that is needed for setting the tool.

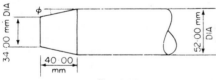

Fig. 1.18

2. Fig. 1.19 shows a groove for a vee-belt. Calculate the distance W.
3. A triangular template is shown in Fig. 1.20. Calculate the distance AB.

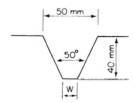

Fig. 1.19

Chapter 1.4 Trigonometry 17

4. Twelve holes are equally spaced on a pitch circle 450 mm diameter. Calculate the length of chord joining the centres of two adjacent holes.
5. A hexagonal nut is 120 mm across flats. Calculate the distance across corners.

The reciprocal ratios. In addition to the sine, cosine and tangent there are three reciprocal ratios which are defined as follows:

$$\text{cosecant } A = \frac{1}{\sin A} \text{ (called } cosec \text{ for short)}$$

$$\text{secant } A = \frac{1}{\cos A} \text{ (called } sec \text{ for short)}$$

$$\text{cotangent } A = \frac{1}{\tan A} \text{ (called } cot \text{ for short)}$$

It often helps to simplify calculations if the unknown length in a triangle trigonometry problem is made the numerator of the trig ratio, as shown in the following problems.

Examples

1. Find the length of the side x in Fig. 1.21.

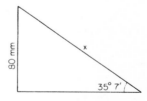

Fig. 1.21

$$\frac{x}{80} = \text{cosec } 35° \ 7'$$

$$x = 80 \times \text{cosec } 35° \ 7' = 80 \times 1·738 = 139·04 \text{ mm}$$

2. Find the length of the side x in Fig. 1.22.

$$\frac{x}{30} = \text{sec } 64°$$

$$x = 30 \times \text{sec } 64° = 30 \times 2·2812 = 68·44 \text{ mm}$$

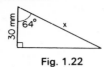

Fig. 1.22

3. Find the length of the side x in Fig. 1.23.

Fig. 1.23

$$\frac{x}{60} = \cot 40°$$

$$x = 60 \times \cot 40° = 60 \times 1\cdot1918 = 71\cdot51 \text{ mm}$$

Exercise 1.12
1. From the tables find the following:
   (a) cosec 39° 27′   (b) cosec 67° 23′
   (c) sec 11° 7′   (d) sec 49° 28′
   (e) cot 37° 49′   (f) cot 74° 11′
2. From the tables find the angle θ if
   (a) cosec θ = 1·3527   (b) sec θ = 1·852
   (c) cot θ = 0·4917
3. Find the lengths of the sides marked x in Fig. 1.24.

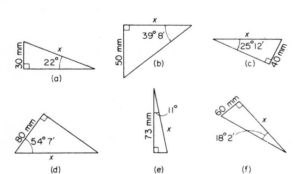

Fig. 1.24

4. By using cosec, sec or cot find the angles marked θ in Fig. 1.25.

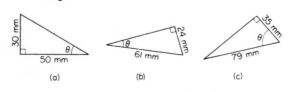

Fig. 1.25

5. Calculate the length which is marked x in Fig. 1.26.

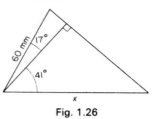

Fig. 1.26

6. Calculate the angle θ in Fig. 1.27.

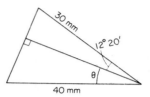

Fig. 1.27

7. The height of an isosceles triangle is 105 mm and each of the equal angles is 39°. Find the lengths of the equal sides.
8. The chord of a circle is 90 mm long and it subtends an angle of 71° at the centre. Calculate the radius of the circle.
9. Twenty holes are equally spaced around the circumference of a circle. If the distance between the centres of two adjacent holes, measured along a chord, is 34·00 mm what is the diameter of the pitch circle?

Logarithms of the trigonometrical ratios. The method of looking up the log of a trig ratio is identical to that used in looking up the ratio itself.

Examples.
1. Find the value of 12·36 × sin 54° 23′.

Number	log
12·63	1·1014
sin 54° 23′	$\overline{1}$·9100
Answer = 10·27	1·0114

2. Find the angle A given that $\cos A = \dfrac{19\cdot87}{26\cdot35}$

Number	log
19·87	1·2983
26·35	1·4208
cos A	$\overline{1}$·8775

The angle A is found directly from the log cos table:
$$A = 41° 2′$$

3. If $b = \dfrac{c(\sin B)}{\sin C}$ find b when $c = 19\cdot28$, $B = 61°$ and $C = 22° 7′$.

Number	log
19·28	1·2851
sin 61°	$\bar{1}$·9418
$c(\sin B)$	1·2269
sin 22° 7′	$\bar{1}$·5757
Answer = 44·79	1·6512

Accuracy of the trig tables. When an angle is smaller than 4° the sine and tangent of the angle are rather inaccurate. If an accurate value for these ratios is required then seven figure tables should be used. For small angles, up to about 6°, the tangent of the angle is very nearly the same as the sine of the angle.

Exercise 1.13
1. Using the tables find the following:
   (a) log sin 28° 33′
   (c) log cos 8° 2′ (d) log cos 24° 15′
   (e) log tan 44° 31′ (f) log tan 7° 5′
   (g) log cosec 71° 10′ (h) log cosec 8° 9′
   (i) log sec 11° 24′ (j) log sec 29° 3′
   (k) log cot 40° 7′ (l) log cot 18° 29′

2. From the tables find the following:
   (a) If log cos A = $\bar{1}$·7357 find the angle A.
   (b) If log sin A = $\bar{1}$·5813 find the angle A.
   (c) If log tan B = 0·5755 find the angle B.
   (d) If log sin ϕ = $\bar{1}$·3069 find the angle ϕ.
   (e) If log cos θ = $\bar{1}$·2381 find the angle θ.
   (f) If log tan α = $\bar{1}$·5569 find the angle α.

3. By using logs find the following:
   (a) If $\cos A = \dfrac{19\cdot26}{27\cdot58}$ find the angle A.

   (b) If $\sin B = \dfrac{11\cdot23}{35\cdot35}$ find the angle B.

   (c) If $\tan\theta = \dfrac{28\cdot13}{17\cdot57}$ find the angle θ.

4. By using logs find the following:
   (a) If $a = \dfrac{81\cdot6\times\sin 43°\ 27′}{\sin 37°\ 11′}$ find a.

   (b) If $\sin C = \dfrac{32\cdot3\times\sin 29°}{51\cdot7}$ find the angle C.

Co-ordinate dimensions. It is often convenient to position holes and other features of a component by means of co-ordinate dimensions, in which case the notation of Fig. 1.6 is used.

Example. Find the co-ordinate dimensions for the three holes shown in Fig. 1.28 relative to the axes OX and OY. The holes lie on a 400 mm pitch circle diameter.

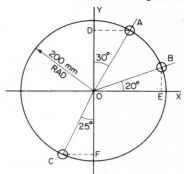

Fig. 1.28

In △ODA,
$$\frac{AD}{OA} = \sin 30°\ \text{or}\ AD = OA\times\sin 30° = 200\times0\cdot5000$$
$$= 100\ \text{mm}$$

$$\frac{OD}{OA} = \cos 30°\ \text{or}\ OD = OA\times\cos 30° = 200\times0\cdot8660$$
$$= 173\cdot2\ \text{mm}$$

Hence for hole A: x co-ordinate = 100 mm
y co-ordinate = 173·2 mm

In △OBE,
$$\frac{EB}{OB} = \sin 20°\ EB = OB\times\sin 20° = 200\times0\cdot3420$$
$$= 68\cdot40\ \text{mm}$$

$$\frac{OE}{OB} = \cos 20°\ \text{or}\ OE = OB\times\cos 20° = 200\times0\cdot9397$$
$$= 187\cdot94\ \text{mm}$$

Hence for hole B: x co-ordinate = 187·94 mm
y co-ordinate = 68·40 mm

In △OFC,
$$\frac{CF}{OC} = \sin 25°\ \text{or}\ CF = OC\times\sin 25° = 200\times0\cdot4226$$
$$= 84\cdot52\ \text{mm}$$

$$\frac{OF}{OC} = \cos 25°\ \text{or}\ OF = OC\times\cos 25° = 200\times0\cdot9063$$
$$= 181\cdot26\ \text{mm}$$

Hence for hole C: x co-ordinate = −84·52 mm
y co-ordinate = −181·26 mm

The results are best displayed in tabular form as shown below:

Hole	x dimension (mm)	y dimension (mm)
A	100·00	173·20
B	187·94	68·40
C	−84·52	−181·26

Exercise 1.14
1. Fig. 1.29 shows five equally spaced holes on a 200 mm pitch circle. Calculate their co-ordinate dimensions relative to OX and OY.

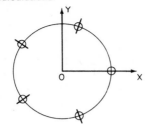

Fig. 1.29

2. Find the co-ordinate hole dimensions for the two holes shown in Fig. 1.30.

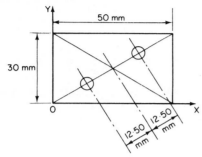

Fig. 1.30

3. Calculate the co-ordinate hole dimensions for the two holes shown in Fig. 1.31.

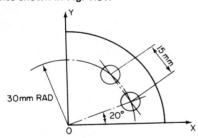

Fig. 1.31

4. Four holes are equally spaced as shown in Fig. 1.32. Find their co-ordinate dimensions relative to the axes OX and OY.

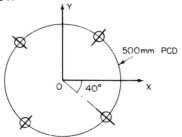

Fig. 1.32

Applications of trigonometry to measurement problems. Many measurement problems can be solved by using a combination of trigonometry and geometry. The most important geometrical theorems are those given below.

A tangent to a circle is at right-angles to a radius drawn from the point of tangency (Fig. 1.33).

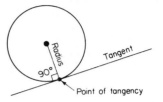

Fig. 1.33

If from a point outside a circle (Fig. 1.34) tangents are drawn to the circle then the two tangents are equal in length. They also make equal angles with the chord joining the points of tangency. The line drawn from the point where the tangents meet to the centre of the circle bisects the angle between the two tangents.

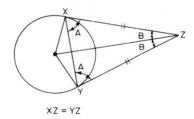

XZ = YZ

Fig. 1.34

If two circles touch internally or externally at one point then the line that passes through their centres also passes through the point of tangency (Fig. 1.35).

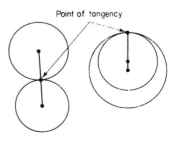

Fig. 1.35

Examples

1. A taper piece is being checked using precision rollers and slip gauges as shown in Fig. 1.36. Find (a) the taper angle and (b) the bottom diameter of the taper piece.

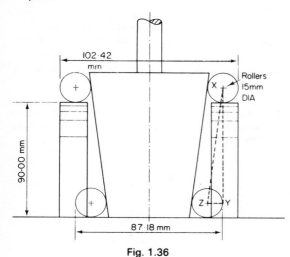

Fig. 1.36

(a) In order to find the angle of taper △XYZ (Fig. 1.37) is used.

$$XY = 90 \cdot 00 \text{ mm}$$

$$ZY = \frac{102 \cdot 42 - 87 \cdot 18}{2} = \frac{15 \cdot 24}{2} = 7 \cdot 62 \text{ mm}$$

$$\angle XYZ = 90°$$

$$\angle ZXY = \theta°$$

$$\tan \theta° = \frac{7 \cdot 62}{90 \cdot 00} = 0 \cdot 0847$$

$$\therefore \theta = 4° \ 50'$$

Angle of taper $= 2 \times \theta = 2 \times 4° \ 50' = 9° \ 40'.$

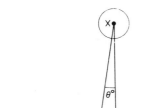

Fig. 1.37

(b) In order to find the bottom diameter of the taper piece we must first calculate dimension EF (Fig. 1.38) since FE and FG are tangents to the roller, BF bisects the angle EFG.

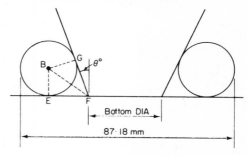

Fig. 1.38

$$\angle EFG = 90° - \theta° = 90° - 4° \ 50' = 85° \ 10'$$
$$\angle EFB = \tfrac{1}{2} \times \angle EFG = \tfrac{1}{2} \times 85° \ 10' = 42° \ 35'$$
$$\frac{EF}{EB} = \cot \angle EFB \text{ or } EF = EB \times \cot \angle EFB$$
$$= 7 \cdot 5 \times \cot 42° \ 35'$$
$$= 7 \cdot 5 \times 1 \cdot 0881$$
$$= 8 \cdot 161 \text{ mm}$$

bottom diameter $= 87 \cdot 18 - 2 \times EF - 2 \times$ roller radius
$$= 87 \cdot 18 - 16 \cdot 322 - 15 \cdot 00$$
$$= 55 \cdot 858 \text{ mm}$$

2. A tapered hole was inspected by using precision balls as shown in Fig. 1.39. The diameters of the balls were 25 mm and 20 mm respectively and the measurements shown in the diagram were obtained. Calculate the included angle of the taper and the diameter of the hole at the top face.

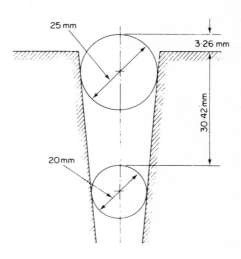

Fig. 1.39

In Fig. 1.40, A and B are the centres of the balls and C and D are the points where the balls touch the side of the hole. Since the angle between a tangent and a radius is a right angle

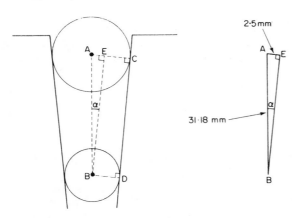

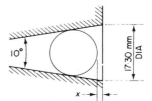

Fig. 1.40

$$\angle ACD = \angle BDC = 90°$$

Draw BE parallel to CD then in $\triangle ABE$

$$AE = \frac{25-20}{2} = 2·5 \text{ mm}$$

$$AB = 30·42+10+3·26-12·5 = 31·18$$

$$\sin \alpha = \frac{AE}{AB} = \frac{2·5}{31·18}$$

$$\alpha = 4° \, 36'$$

Included angle of taper = $2\alpha = 2×4° \, 36' = 9° \, 12'$.

In order to calculate the diameter at the top of the hole the construction shown in Fig. 1.41 is required. From $\triangle AFC$ we can calculate AF and from $\triangle FGH$ we can calculate HG. The radius at the top of the hole is AF+HG. In $\triangle AFC$,

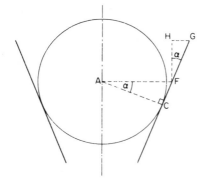

Fig. 1.41

AC = radius of the ball = 12·5 mm

$$\frac{AF}{AC} = \sec \alpha \text{ or } AF = AC × \sec 4° \, 36'$$
$$= 12·5×1·0032 = 12·54 \text{ mm}$$

In $\triangle FGH$, HF = 12·5−3·26 = 9·24 mm

$$\frac{HG}{HF} = \tan \alpha \text{ or } HG = HF × \tan 4° \, 36'$$
$$= 9·24×0·0805 = 0·74 \text{ mm}$$

Radius at the top of the hole = AF+HG = 12·54+0·74
$$= 13·28 \text{ mm}$$

Diameter at the top of the hole = 2×13·28 = 26·56 mm

Exercise 1.15

1. A steel precision ball 15 mm diameter is used to check a taper hole, a section of which is shown in Fig. 1.42. If the taper is correct what is the dimension *x*?

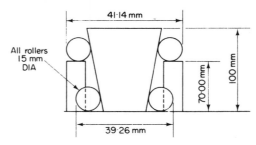

Fig. 1.42

2. A taper plug gauge is being checked by means of reference rollers and balls as shown in Fig. 1.43. Find the included angle of the taper and the top and bottom diameters of the gauge.

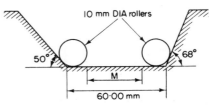

Fig. 1.43

3. Calculate the dimension *M* that is needed for checking the groove shown in Fig. 1.44.

Fig. 1.44

4. Fig. 1.45 shows the dimensions obtained when checking a tapered hole. Find the included angle of taper and the top diameter *d* of the hole.

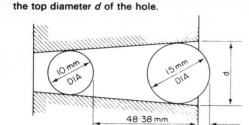

Fig. 1.45

5. Fig. 1.46 shows part of a bracket. Find the dimension *W* that is needed for laying out the bracket.

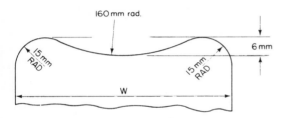

Fig. 1.46

6. Fig. 1.47 shows a template. The dimension *x* is required in order to lay-out the work. Calculate its value.

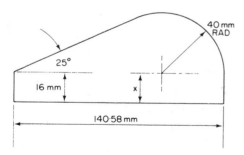

Fig. 1.47

Checking profiles. Profiles are often checked by using the properties of chords. The most important geometrical theorems are:

If the diameter of a circle is at right angles to a chord, then it divides the chord into two equal parts (Fig. 1.48).

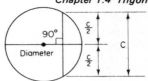

Fig. 1.48

If two chords intersect in a circle the product of the segments of one chord is equal to the product of the segments of the other (Fig. 1.49).

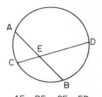

AE x BE = CE x ED

Fig. 1.49

If a triangle is inscribed in a semi-circle the angle opposite the diameter is a right angle (Fig. 1.50).

Fig. 1.50

Examples

1. Fig. 1.51 shows a method of checking the radius of a profile when actual measurement is impossible. If $H = 25 \cdot 00$ mm and $L = 80 \cdot 00$ mm find the radius of the profile.

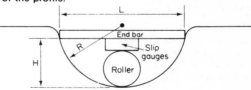

Fig. 1.51

Using the theorem of intersecting chords (Fig. 1.52):

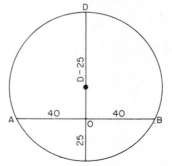

Fig. 1.52

$$OC \times OD = AO \times OB$$
$$25 \times (D-25) = 40 \times 40$$
$$25 \times (D-25) = 1600$$
$$D-25 = \frac{1600}{25} = 64$$
$$D = 89 \text{ mm}$$

The radius of the profile is $\frac{89}{2} = 44 \cdot 5$ mm.

2. Fig. 1.53 shows the profile of a template. The distance W is required for checking purposes. What is its value if the template is correct?

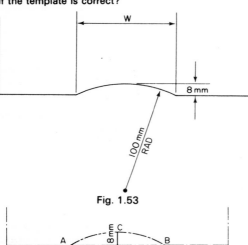

Fig. 1.53

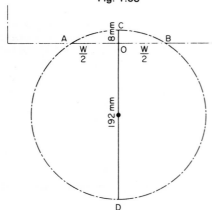

Fig. 1.54

Referring to Fig. 1.54:
$$AO \times OB = OC \times OD$$
$$\frac{W}{2} \times \frac{W}{2} = 8 \times 192$$
$$\frac{W^2}{4} = 1536$$
$$W^2 = 6144$$
$$W = \sqrt{6144}$$
$$W = 78 \cdot 40 \text{ mm}$$

Tangential contacts. When laying out templates, etc., it is often necessary to join up circular arcs and straight lines. Where accuracy is essential, as is the case with optical projection, a calculation such as the one given below is necessary.

Example The profile shown in Fig. 1.55 is to be marked out accurately. In order to spot the centre of the arc the distance M is required. Calculate its value.

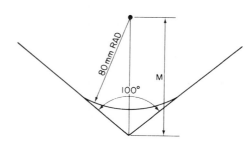

Fig. 1.55

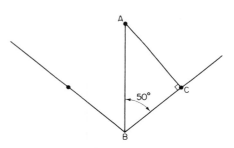

Fig. 1.56

Using Fig. 1.56 and the tangent properties of a circle:
In $\triangle ABC$,
$$\angle ABC = 50°$$
$$\angle ACB = 90°$$
$$AB = M$$
$$AC = 80 \text{ mm}$$
$$\frac{AB}{AC} = \text{cosec } \angle ABC$$
$$\frac{M}{80} = \text{cosec } 50°$$
$$M = 80 \times 1 \cdot 3054 = 104 \cdot 43 \text{ mm}$$

Exercise 1.16
1. Fig. 1.57 shows the measurements obtained when checking a concave surface. Calculate the radius of the surface.

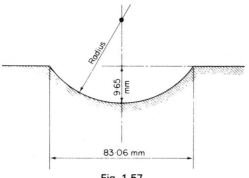

Fig. 1.57

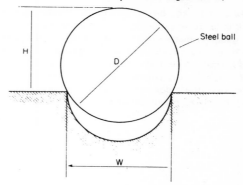

Fig. 1.60

2. When finding the radius of a segment having a rounded edge the set-up shown in Fig. 1.58 is used. If the width of the micrometer frame is 85·60 mm and the reading on the micrometer is 30·48 mm find the radius of the segment.

5. The template shown in Fig. 1.61 has to be marked out accurately. Find the dimension *M* that is needed for this purpose.

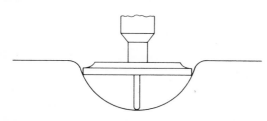

Fig. 1.58

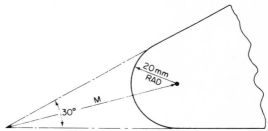

Fig. 1.61

3. Fig. 1.59 shows part of a turned bar. Calculate the dimension *C*.

6. The profile shown in Fig. 1.62 is to be marked out. Calculate the co-ordinates for the contact points P and Q (i.e. calculate *x, y, X* and *Y*) that are needed for this purpose.

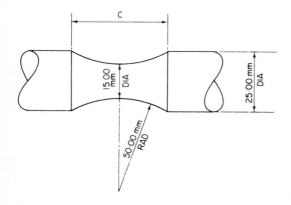

Fig. 1.59

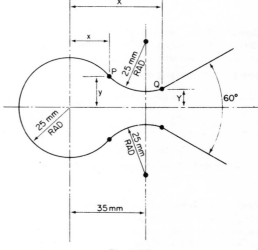

4. Fig. 1.60 shows a method of checking the width of a small concave surface. If *H* = 13·72 mm and *D* = 15·00 mm find *W*.

Fig. 1.62

Measuring large bores. One way of measuring a large bore is to use a pin gauge as shown in Fig. 1.63. The pin gauge is made slightly smaller than the diameter of the bore to be measured. When the gauge is held at one end a small amount of rocking is obtained at the other end. Knowing the length L of the pin gauge and, by measuring, the amount of rock 2ω, the diameter D of the bore can be calculated by using the formula

$$D = L + \frac{\omega^2}{2L}$$

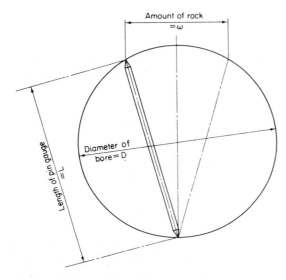

Fig. 1.63   Using a pin gauge to measure the diameter of a large bore

Example. When using a pin gauge whose length is 400 mm the amount of rock obtained was 5 mm. Calculate the diameter of the bore being gauged.

Here L = 400 mm and ω = 2·5 mm

$$D = L + \frac{\omega^2}{2L} = 400 + \frac{2·5^2}{2 \times 400}$$

$$= 400 + \frac{6·25}{800} = 400 + 0·0078$$

$$= 400·078 \text{ mm}$$

Hence the diameter of the bore is 400·078 mm.

Measuring large radii. A quick method of measuring a large radius is shown in Fig. 1.64. The vee gauge is placed on the workpiece and the perpendicular distance from the workpiece to the vee is measured by using a micrometer head. It can be shown that

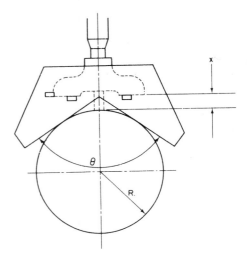

Fig. 1.64   Measuring work with a large radius by means of a vee gauge and a micrometer head.

$$R = x\left(\frac{1}{\text{cosec } \theta/2 - 1}\right)$$

where R = radius of work (millimetres)

$\quad x$ = distance from workpiece to vee (millimetres)

$\quad \theta$ = included angle of the vee (degrees)

Since θ will always have the same value for any given vee the value of $\dfrac{1}{\text{cosec } \theta/2 - 1}$ can be calculated and marked on the gauge. Thus, when θ = 120°,

$$\frac{1}{\text{cosec } \theta/2 - 1} = \frac{1}{\text{cosec } 60° - 1} = \frac{1}{1·1547 - 1}$$

$$= \frac{1}{0·1547} = 6·465$$

Hence for this gauge, $R = 6·465x$

Example. A vee gauge with a 120° vee is used to measure the radius of round work. If the reading on the micrometer is 15·46 mm what is the radius of the work?

It has been shown above that the radius of the work is given by

$$R = 6·465 \, x$$

Thus when x = 15·46 mm

$$R = 6·465 \times 15·46 = 99·93 \text{ mm}$$

Exercise 1.17
1. When using a pin gauge whose length is 500 mm the amount of rock obtained was 8 mm. Calculate the diameter of the bore being gauged.
2. A pin gauge is 420 mm long and when rocked in a bore the amount of rock obtained was 5·3 mm. What is the diameter of the bore?

3. In order to find the radius of a large arc a vee-gauge with an angle of 150° was used. If the distance from the vertex of the vee to the arc is 18·16 mm what is the radius of the arc?

4. A vee-gauge with a 90° vee is used to measure the radius of round work. If the radius is supposed to be 400 mm find the distance from the vertex of the vee assuming that the radius is correct.

The sine rule. This rule is used when a triangle is *not* right-angled. Using the notation of Fig. 1.65 the rule states:

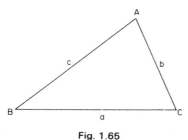

Fig. 1.65

$$\frac{a}{\sin A} = \frac{b}{\sin B} = \frac{c}{\sin C}$$

Example. In △ABC (Fig. 1.66) $\angle A = 58°$, $\angle C = 72° \, 20'$ and $b = 52·3$ mm. Calculate the lengths of the sides a and c.

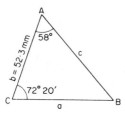

Fig. 1.66

Since $\angle A + \angle B + \angle C = 180°$

$$\angle B = 180° - 58° - 72° \, 20'$$
$$\angle B = 49° \, 40'$$

The sine rule states:

$$\frac{a}{\sin A} = \frac{b}{\sin B}$$
$$a = \frac{b \sin A}{\sin B}$$
$$= \frac{52·3 \sin 58°}{\sin 49° \, 40'}$$
$$a = 58·19 \text{ mm}$$

Number	log
52·3	1·7185
sin 58°	$\bar{1}$·9284
	1·6469
sin 49° 40'	$\bar{1}$·8821
58·19 ←	1·7648

Also, $\dfrac{c}{\sin C} = \dfrac{b}{\sin B}$

$$c = \frac{b \sin C}{\sin B}$$
$$= \frac{52·3 \sin 72° \, 20'}{\sin 49° \, 40'}$$
$$c = 65·37 \text{ mm}$$

Number	Log
52·3	1·7185
sin 72° 20'	$\bar{1}$·9790
	1·6975
sin 49° 40'	$\bar{1}$·8821
65·37 ←	1·8154

Angles greater than 90°. The method shown above is satisfactory for acute-angled triangles but if the triangle contains an obtuse angle then we have to find the sine of an angle which is greater than 90°. This is easily done if we remember the construction of Fig. 1.67.

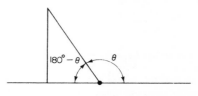

Fig. 1.67

When θ is greater than 90°

$$\sin \theta = \sin (180° - \theta)$$

Examples.
1. $\sin 150° = \sin (180° - 150°) = \sin 30° = 0·5000$
2. $\sin 139°4' = \sin (180° - 139°4') = \sin 40°56' = 0·655 \, 1$.

Example. Three holes are positioned as shown in Fig. 1.68. Calculate the centre distances from A to B.

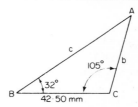

Fig. 1.68

$$\angle A = 180° - 32° - 105° = 43°$$

By the same rule

$$\frac{c}{\sin C} = \frac{a}{\sin A}$$
$$c = \frac{a \sin c}{\sin A}$$
$$= \frac{42·50 \times \sin 105°}{\sin 43°}$$
$$= \frac{42·50 \times \sin 75°}{\sin 43°}$$
$$= 60·19 \text{ mm}$$

Number	log
42·50	1·6284
sin 75°	$\bar{1}$·9849
	1·6133
sin 43°	$\bar{1}$·8338
60·19	1·7795

The diameter of the circumscribing circle of a triangle. Using the notation on the diagram (Fig. 1.69), the same rule states

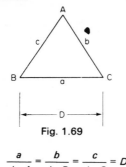

Fig. 1.69

$$\frac{a}{\sin A} = \frac{b}{\sin B} = \frac{c}{\sin C} = D$$

This rule is often used to find the pitch circle diameter of a ring of holes.

Example. In Fig. 1.70, three holes are positioned by the angle and dimensions given. Find the pitch circle diameter.

Fig. 1.70

We are given $\angle B = 41°$ and $b = 112.5$ mm

$$\frac{b}{\sin B} = D$$

$$\therefore D = \frac{112.5}{\sin 41°} = 171.5 \text{ mm}$$

Hence the pitch circle diameter is 171·5 mm.

Exercise 1.18

1. Solve the following triangles ABC given
    (a) $A = 75°$ $B = 34°$ $a = 102$ mm
    (b) $C = 61°$ $B = 71°$ $b = 91$ mm
    (c) $A = 19°$ $C = 105°$ $c = 111$ mm
    (d) $B = 116°$ $C = 18°$ $a = 170$ mm
    (e) $A = 36°$ $B = 77°$ $b = 250$ mm
    (f) $A = 49°11'$ $B = 67°17'$ $c = 111.2$ mm
    (g) $A = 17°15'$ $C = 27°7'$ $b = 22.15$ mm
    (h) $A = 77°3'$ $C = 21°3'$ $a = 97.93$ mm
    (i) $B = 115°4'$ $C = 11°17'$ $c = 51.62$ mm
    (j) $a = 92.17$ mm $b = 71.52$ mm $A = 105°4'$
2. In the slider crank mechanism shown in Fig. 1.71 calculate the distance of the slider from the top of its stroke (i.e. find the distance *x*).

Fig. 1.71

3. The mechanism shown in Fig. 1.72 comprises two sliders. If $\angle ABO = 25°$ and $\angle AOB = 120°$ find the length AO.

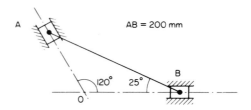

Fig. 1.72

4. Three holes lie on a pitch circle as shown in Fig. 1.73. Calculate the pitch circle diameter.

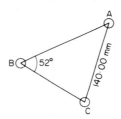

Fig. 1.73

5. In Fig. 1.74, A and B represent the centres of two gears. An idler is placed with its centre at C. Find the distances AB and AC.

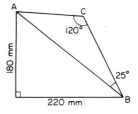

Fig. 1.74

6. Three holes A, B, and C (Fig. 1.75) are positioned on a pitch circle whose diameter is 200 mm. Calculate the distances AB, BC and AC that are needed for checking purposes.

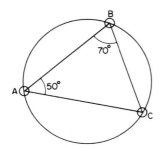

Fig. 1.75

Formulae containing trig ratios. There are many formulae which contain one or more of the trig ratios. The methods used in evaluating them are shown in the following examples.

Examples

1. A formula used when checking a screw thread is
$$M = D+d(1+\text{cosec } \theta)-\frac{p}{2}\cot \theta$$
Find the value of M when $D = 46.75$, $p = 5$, $d = 3$ and $\theta = 30°$.
Putting the given values into the formula
$$M = 46.67+3\times(1+\text{cosec } 30°)-\frac{5}{2}\times\cot 30°$$
$$= 46.75+3\times(1+2)-2.5\times1.7321$$
$$= 46.75+9-4.33$$
$$= 51.42$$

2. When determining the maximum width of a tapered hole the following formula is used:
$$M = 2\tan\frac{A}{2}\left[D_2-D_1+R(1+\text{cosec}\frac{A}{2})\right]$$
Find M when $A = 20°$, $D_1 = 40$ mm, $D_2 = 35$ mm and $R = 10$ mm.
Substituting the given values in the formula,
$$M = 2\times\tan 10°\times[40-35+10\times(1+\text{cosec } 10°)]$$
$$= 2\times0.1763\times[5+10\times(1+5.759)]$$
$$= 0.3526\times[5+10\times6.759]$$
$$= 0.3526\times[5+67.59]$$
$$= 0.3526\times72.59$$
$$= 25.59$$

Squares of the trig ratios. The square of sin A is written sin² A and the square of tan θ is written tan² θ, etc.

Examples

1. $\tan^2 20° = (0.3640)^2 = 0.1325$
2. $\sec^2 35° = (1.2208)^2 = 1.490$

Exercise 1.19

1. The lead of a helical gear is given by the formula $L = \dfrac{\pi mT\sec \alpha}{\tan \alpha}$. Find the values of L when $m = 4$, $\alpha = 30°$ and $T = 26$.

2. When using a gear tooth vernier the formula
$$h = m+\frac{mT}{2}\left(1-\cos\frac{90}{T}\right)$$
is used. Calculate the value of h when $m = 2.5$ and $T = 32$.

3. Another formula used in connection with a gear tooth vernier is $h = m\left[1-\dfrac{\pi}{4}\cos \phi \sin \phi\right]$. Find the value of h when $m = 5$, $T = 25$ and $\phi = 20°$.

4. The following formula is used when checking a taper:
$D = C+2H\tan\dfrac{\alpha}{2}-d\sec\dfrac{\alpha}{2}$. Find D when $H = 15$, $C = 50$, $\alpha = 30°$ and $d = 12$.

5. When measuring a large radius the formula
$R = \dfrac{h}{\text{cosec}\dfrac{\theta}{2}-1}$ is used. Find R when $\theta = 120°$ and $h = 11.75$.

6. Find the values of: (a) sin² 40° (b) cos² 27° (c) tan² 80° (d) cot² 22° (e) sec² 18° (f) cosec² 71°.

7. If $P = (1+\cot^2 A)\sin^2 A$ find P when $A = 40°$.

8. In a measuring problem the formula
$$L = h\tan \theta+r\tan\left(45°-\frac{\theta}{2}\right)$$
is used. Find L when $h = 30$, $r = 12$ and $\theta = 32°$.

Part 2 Science

2.1 SI units

Unit of mass. The mass of a body is the quantity of matter that it contains. In SI the standard unit of mass is the kilogram (kg) but very small masses may be measured in grams (g).

Unit of force. Forces are measured in newtons (N), kilonewtons (kN) or meganewtons (MN).

$$1000 \text{ N} = 1 \text{ kN}$$
$$1\,000\,000 \text{ N} = 1 \text{ MN}$$

Weight. The weight of a body is the force that the earth exerts upon it. Weight is therefore a force and it is measured in newtons.

Relationship between mass and weight.

Weight (newtons) = 9·81 × mass (kilograms)

Example. A forging has a mass of 200 kg. What is the weight of the forging?

Weight = 9·81 × mass = 9.81 × 200 = 1962 N

Unit of length. The standard unit of length is the metre (m). Large distances are measured in kilometres (km) but small distances are measured in millimetres (mm) or micrometres (μm).

$$1\,000\,000 \; \mu\text{m} = 1 \text{ m}$$
$$1\,000 \text{ mm} = 1 \text{ m}$$
$$1 \text{ km} = 1000 \text{ m}$$

Unit of time. The standard unit of time is the second (s) but time may also be measured in minutes (min) or hours (h).

Unit of temperature. The standard unit of temperature is the kelvin (K). Kelvins are absolute degrees Celsius (Celsius being the SI name for the Centigrade temperature scale) that is, the temperature is measured from absolute zero which is −273°C. Thus

kelvins = degrees Celsius + 273

Examples.
1. A temperature is 200 °C. How many kelvins is this?

kelvins = 200 + 273 = 473 K

(Note that we do not write 473°K but only 473K)

2. A temperature is 630K. What is this in °C?

°C = kelvins − 273 = 630 − 273 = 357 °C

Unit of pressure. Pressure is defined as force per unit area. That is

$$pressure = \frac{force}{area}$$

Since the unit of force is the newton and the unit of area is the square metre (m²) the standard unit of pressure is newtons per square metre (N/m²). However large pressures may be measured in kilonewtons per square metre (kN/m²) or meganewtons per square metre (MN/m²).

Examples.
1. A force of 2000 N acts uniformly over an area of 4 m². What is the pressure?

$$pressure = \frac{force}{area} = \frac{2000}{4} = 500 \text{ N/m}^2$$

2. A force of 200 N acts uniformly on an area of 40 mm². What is the pressure?

$$1 \text{ mm} = \frac{1}{1000} \text{ m}$$

$$1 \text{ mm}^2 = \frac{1}{1000 \times 1000} = \frac{1}{1\,000\,000} \text{ m}^2$$

$$pressure = \frac{force}{area} = \frac{200}{\dfrac{40}{1\,000\,000}}$$

$$= \frac{200 \times 1\,000\,000}{40}$$

$$= 5\,000\,000 \text{ N/m}^2 \text{ or } 5 \text{ MN/m}^2$$

(It is worth while remembering that
$$1 \text{ N/mm}^2 = 1 \text{ MN/m}^2)$$

Another unit of pressure which is used is the *bar*. This is a useful unit because 1 bar very nearly equals atmospheric pressure. The correct relationship is

Atmospheric pressure = 1·013 bar

To convert bars into newtons per square metre we use the relationship

$$1 \text{ bar} = 100\,000 \text{ N/m}^2$$

Examples.
1. A pressure is stated to be 12 bars. Express this pressure in newtons per square metre.

12 bars = 12 × 100 000 = 1 200 000 N/m² or
1·2 MN/m²

2. A pressure is 7 MN/m². Express this in bars.

7 MN/m² = 7 000 000 N/m²

$$= \frac{7\,000\,000}{100\,000} = 70 \text{ bars}$$

Exercise 2.1

1. A casting has a mass of 800 kg. Calculate its weight.
2. A billet of steel weighs 1800 N. What is the mass of the billet?
3. A temperature is 300°C. Express this temperature in kelvins.
4. A temperature is found to be 536 K. What is the temperature in degrees Celsius?
5. A force of 4000 N acts uniformly over an area of 8 mm². What is the pressure on the area?
6. A force of 6000 N acts uniformly on an area of 12 cm². Calculate the pressure on the area (a) in N/cm² (b) in N/mm² and (c) in KN/m².
7. A disc whose area is 4 cm² has a pressure of 8 MN/m² acting on it. What is the total force acting on the disc?
8. A force of 1200 N acts on a piston whose diameter is 25 mm. Calculate the pressure in the cylinder.
9. A certain pressure is stated to be 15·8 bars. What is this pressure in MN/m²?
10. The cylinder of a jack used to operate a machine tool has a diameter of 100 mm. The operating pressure is 120 bars. Calculate the force acting on the full diameter of the piston.
11. The ram of a hydraulic press has a diameter of 2000 mm. If the pressure in the cylinder is 180 bars calculate the force acting on the ram.
12. In a hydraulic jack the piston is 30 mm dia. and the piston rod is 10 mm dia. If the pressure in the cylinder is 80 bars calculate the force acting (a) on the forward stroke (b) on the return stroke.

Unit of work.

Work done = force × distance moved in the direction of the force.

Since the unit of force is the newton and the unit of distance is the metre the unit of work done is the newton metre. However the standard unit of work is the joule (J) such that

$$1 \text{ joule} = 1 \text{ newton metre}$$

Example. The force at the point of a shaping machine when it is cutting is 1500 N. If the length of the stroke is 120 mm how much work is done in one cutting stroke?.

Work done = force × distance

$$= 1500 \times \frac{120}{1000} = 180 \text{ Nm or } 180 \text{ J}$$

Unit of power. Power is the rate of doing work that is,

$$power = \frac{work \ done}{time \ taken}$$

Since the standard unit of work is the joule and the unit of time is the second, power is measured in joules per second (J/s). However power is always measured in watts (W) and

$$1 \text{ watt} = 1 \text{ joule per second}$$

Example. When shaping, the force on the cutting tool was 1800 N and the length of stroke was 200 mm. If a single cutting stroke takes 2 s find the power used.

Work done = force × distance

$$= 1800 \times \frac{200}{1000} = 360 \text{ J}$$

Time taken = 2 s

$$\text{Power used} = \frac{\text{Work done}}{\text{time taken}} = \frac{360}{2} = 180 \text{ J/s or } 180 \text{ W}$$

Power and cutting speed. When the cutting speed and the force at the tool point are known the power may be calculated from

power (watts) = force (newtons) × cutting speed (metres per second)

Example. During planing the force at the tool point is 5000 N and the cutting speed is 24 m/min. Calculate the power used in cutting.

Power = force × cutting speed

$$= 5000 \times \frac{24}{60} = 2000 \text{ W or 2 kW}$$

Power and torque. Many machine tools have revolving spindles or revolving cutters. In such cases the power used may be calculated from

power (watts) = 2π × torque (newton metres) × rotational speed (revolutions per second)

It will be remembered that torque is another name for 'turning moment' and is calculated as follows:

torque (newton metres) = force (newtons) × distance (metres)

Example. When cutting a bar 80 mm diameter the force at the cutting tool is 4000 N (Fig. 2.1). If the workpiece revolves at 90 rev/min calculate the power used.

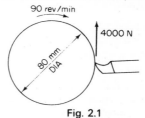

Fig. 2.1

Torque = force × radius = $4000 \times \frac{40}{1000} = 160$ Nm.

Rotational speed = $\frac{90}{60} = 1 \cdot 5$ rev/s

Power used = $2\pi \times 160 \times 1 \cdot 5 = 1507$ W or $1 \cdot 507$ kW.

Hydraulic power. The power developed in a hydraulic system depends upon the pressure of the fluid and the quantity of fluid flowing. Thus,

power (watts) = pressure (newtons per square metre) × quantity flowing (cubic metres per second)

Example. A pump delivers 1200 litres per minute at a pressure of 350 kN/m². What power does the pump use?

Since 1 litre = 1000 cm³ = $\dfrac{1000}{100 \times 100 \times 100}$ = $\dfrac{1}{1000}$ m³

Quantity of fluid flowing = $1200 \times \dfrac{1}{1000} \times \dfrac{1}{60}$ = 0·02 m³/s

Pressure of fluid = 350 kN/m² = 350 000 N/m².

Power = pressure × quantity flowing = 350 000 × 0·02
$$= 7000 \text{ W or } 7 \text{ kW}$$

Loss of power due to friction. The coefficient of friction is defined as

$$\mu = \frac{F}{W}$$

where F = sliding friction force (newtons)
  and W = force between the surfaces (newtons) (see Fig. 2.2)

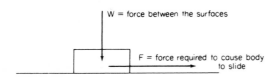

W = force between the surfaces

F = force required to cause body to slide

Fig. 2.2

Example. The table of a planing machine and the work upon it together have a mass of 2000 kg. If the co-efficient of friction between the slides and the table is 0·04 find the least force that will cause the table to move steadily along the slides.

Since $\dfrac{F}{W} = \mu$

then $F = \mu W$

The force between the surfaces is
$$W = 9·81 \times 2000 = 19\ 620 \text{ N}$$
The least force that will cause the table to move is
$$F = \mu W = 0·04 \times 19\ 620 = 784 \text{ N}$$

When part of a machine tool operates by sliding (e.g. the ram of a shaping machine) power is lost in overcoming the frictional resistance.

Example. The ram of a shaping machine has a mass of 200 kg and the coefficient of friction between the ram and the slides is 0·05. If the ram is moving at a speed of 10 m/min find the power lost in friction.
The force between the surfaces is

$$W = 9·81 \times 200 = 1962 \text{ N}$$

Force needed to move the ram against the frictional resistance is

$$F = \mu W = 0·05 \times 1962 = 98·1 \text{ N}$$

The power lost in overcoming friction is
P = force needed to overcome friction × speed of ram

$$= 98·1 \times \frac{10}{60} = 16·4 \text{ W}$$

(Note that the speed of the ram must be stated in metres per second.)
  Power is also lost when a shaft rotates in a bearing.

Example. A shaft has a diameter of 200 mm and the load on the bearing is 20 kN. The coefficient of friction between the shaft and its bearing is 0·05 and the shaft revolves at 600 rev/min. Calculate the power lost in friction.
Referring to Fig. 2.3:
Force needed to overcome friction is
$$F = \mu W = 0·05 \times 20\ 000 = 1000 \text{ N}$$
The friction torque is
$$T = F \times \text{radius of shaft}$$
$$= 1000 \times \frac{100}{1000} = 100 \text{ Nm}$$
The power lost in friction is
$$P = 2\pi \times T \times \text{rotational speed (in rev/s)}$$
$$= 2\pi \times 100 \times \frac{600}{60}$$
$$= 6284 \text{ W or } 6·284 \text{ kW}$$

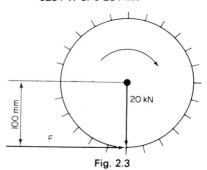

100 mm   20 kN

F

Fig. 2.3

Exercise 2.2

1. The force acting on the point of a shaping machine when it is cutting is 2000 N. If the length of the cut is 150 mm calculate the work done during cutting.
2. A forging has a mass of 400 kg. Calculate the work done in lifting the forging through a height of 8 m.
3. A lifting machine lifts a load of 120 kg through a height of 15 m in 30 s. How much power is used?
4. When milling, the traverse speed of the table is 120 mm/min. If the force required to feed the table is 4400 N calculate the power used in traversing the table.
5. A planing machine cuts metal at a speed of 25 m/min. If the force at the toolpoint is 2000 N calculate the power used in cutting.
6. When turning work whose diameter is 60 mm the tangential force at the toolpoint is 5000 N. If the

workpiece revolves at 200 rev/min calculate the power used in cutting.

7. The torque on a drill is 30 Nm. If it is rotating at 500 rev/min calculate the power used in drilling.

8. A milling cutter, 80 mm dia., revolves at 400 rev/min. If the cutting force at the teeth is 1500 N calculate the power required.

9. A pump delivers oil at a rate of 1·5 m³/min with a pressure of 6 bars. What power does the pump use?

10. The delivery rate to a hydraulic jack is 80 litres per minute with a pressure of 10 MN/m². What power is developed?

11. A hydraulic jack used to operate a milling machine develops a power of 5 kW. If the delivery rate is 100 litres per minute find the pressure in the cylinder.

12. The table of a planing machine has a mass of 5000 kg and the coefficient of friction between the table and the slides is 0·05. If the table speed is 15 m/min calculate the power lost in overcoming friction.

13. The carriage of a lathe has a mass of 400 kg and it is moving at a speed of 80 mm/min. If the co-efficient of friction between the carriage and the slides is 0·06 find the power lost in friction.

14. The load on a bearing is 50 kN. If the shaft is 160 mm dia and rotates at 200 rev/min find the power lost in overcoming friction. (Take the coefficient of friction at the bearing as 0·05.)

2.2   Heat

Effect of heat on solids. When heat is applied to a solid the solid expands.

If a bar of metal is heated it increases in length and when it cools it contracts. The amount of expansion or contraction is calculated from the formula

alteration in length = original length × coefficient of linear expansion × alteration in temperature

Examples.

1. A steel length bar 200 mm long at 20 °C has its temperature raised to 30 °C. If the coefficient of linear expansion of steel is 0·000 011 2 per 1 °C calculate the increase in length of the bar.

   Alteration in temperature = 30°−20° = 10 °C

   Increase in length = 200×0·000 011 2×10 = 0·0224 mm. Although this is only a small amount of expansion it is important when dealing with accurate measuring devices such as length bars.

2. A component is brought into the measuring room and checked against a standard end bar. If the component is at 10 °C and the end bar is at 20 °C and the component has a nominal length of 400 mm find the error in the measurement. (Coefficient of linear expansion of the component 0·000 018 per °C).

Alteration in temperature = 20°−10° = 10 °C
Error in measurement   = 400×0·000 018×10
                        = 0·072 mm

When measuring, it is important that the measuring device and the component are at the standard temperature of 20 °C. When gauging, the temperature does not matter provided (a) the temperature of the component and the gauge are the same and (b) the component and the gauge are made from the same material.

Shrink fits. Collars and other parts that have to be a permanent fit on a shaft are sometimes fitted by shrinking them onto the shaft. The collar is made slightly smaller than the shaft. It is then heated so that it expands sufficiently for it to fit easily over the shaft. On cooling the collar contracts and thus fits tightly.

Example. A steel collar is to be shrunk onto a shaft. The hole in the collar is 199·80 mm at 15 °C and is required to be 200·80 mm when heated so that it fits easily over the shaft. Find the temperature to which the collar must be heated for this diameter to be obtained. (Coefficient of linear expansion is 0·000 012 per °C).
Expansion required on diameter = 200·80−199·80 = 1·00 mm.

Expansion = Original diameter × coefficient of linear expansion × temperature rise

∴ 1·00   = 199·80 × 0·000 012 × temperature rise

$$\text{temperature rise} = \frac{1\cdot00}{199\cdot80 \times 0\cdot000\ 012} = 417$$

Hence required temperature = 417° + 15° = 432 °C

Example. A steel tyre is to fit tightly over a wheel, 2 m diameter, when the tyre is 100 °C above the temperature of the wheel. To what internal diameter must the tyre be turned? (Coefft. of linear expansion is 0·000 011 per °C).

The diameters of the tyre and wheel must be the same when the temperature of the tyre is 100 °C above that of the wheel. Hence at 100°C above the wheel temperature the tyre must have a diameter of 2 m. Let the temperature of the tyre drop to room temperature, i.e. sustain a drop of 100 °C, then

contraction of tyre = original diameter × coefficient of
linear expansion × temperature drop
= 2 × 0·000 011 × 100
= 0·002 2 m or 2·2 mm

Hence internal diameter at room temperature
= 2000 − 2·2 = 1997·8 mm

Coefficient of cubic expansion. A solid such as a block of metal does not expand and contract lengthways only. The height and breadth of the block also expand and contract and hence the volume of the block will increase when heated and decrease when cooled.

The amount that a volume increases or decreases is given by the formula

*alteration in volume = original volume × coefficient of
cubic expansion × alteration in
temperature.*

The coefficient of cubic expansion may be taken as being three times the coefficient of linear expansion. Thus, for steel

Coefficient of linear expansion = 0·000 011 2 per 1 °C
Coefficient of cubic expansion = 3 × 0·000 011 2
= 0·000 033 6 per 1 °C

Example. A block of cast iron is 80 × 60 × 40 mm at 20°C. Calculate the increase in volume of the block when it is heated to a temperature of 80°C. (Coefficient of linear expansion for cast iron is 0·000 010 per 1 °C.)

Volume of block = 80 × 60 × 40 = 192 000 mm³

Coefficient of cubic expansion = 3 × 0·000 010 = 0·000 030 per 1 °C.

Increase in temperature = 80° − 20° = 60 °C

Increase in volume   = 192 000 × 0·000 030 × 60
= 3456 mm³

Exercise 2.3

1. A steel bar 800 mm long has its temperature raised from 15 °C to 220 °C. If the coefficient of linear expansion for steel is 0·000 011 2 per 1 °C calculate the amount the bar expands.

2. A steel plug gauge 25·00 mm diameter at 20 °C is used in an unheated store where the temperature is 12 °C. What is the diameter of the gauge at 12 °C? (Coefficient of linear expansion for steel is 0·000 011 2 per 1 °C.)

3. A component was brought in from an unheated store and checked against a standard end bar. The temperature of the component was 7 °C when the check took place but the end bar was at the standard temperature of 20 °C. If the nominal length of the component was 300 mm calculate the error made in the measurement. (Coefficient of linear expansion of the component is 0·000 020 per 1°C.)

4. A collar having a bore of 120 mm at 15 °C has to be expanded by 0·16 mm to slide over a shaft. Calculate the temperature to which the collar must be heated. (Coefficient of linear expansion for the collar is 0·000 017 per 1 °C.)

5. A steel collar has a bore of 200·00 mm dia. at 20 °C. The bore has to be expanded to 200·30 mm to fit over a shaft. Find the temperature to which the collar must be heated. (Coefficient of linear expansion for steel is 0·000 011 per 1 °C.)

6. A ball race having a nominal diameter of 50 mm at 20 °C has to have its diameter shrunk by 0·05 mm so that it can fit in its housing. To what temperature must the ball race be cooled? (Coefficient of linear expansion for the ball race is 0·000 011 per 1 °C.)

7. A block of aluminium 20 × 30 × 40 mm is heated from 15 °C to 150 °C. If the coefficient of linear expansion for aluminium is 0·000 024 per 1 °C calculate the increase in the volume of the block.

8. A brass cylinder 40 mm diameter and 100 mm long is heated from 20 °C to 220 °C. Find (a) the increase in the diameter of the cylinder and (b) the increase in its volume. Take coefficient of linear expansion as 0·000 020 per 1 °C.

Quantity of heat. The quantity of heat used in heating a body is given by the formula

*quantity of heat (joules) = mass of body (kilograms) ×
specific heat (joules per kilogram °C) × temperature rise
(°C).*

The same formula may be used when a body loses heat but in this case a temperature drop is used instead of a temperature rise.

Examples.

1. A piece of copper having a mass of 20 kg is heated from 20 °C to 450 °C. If its specific heat is 377 J/kg °C calculate the heat gained by the copper.

Rise in temperature = 450° − 20° = 430 °C

Heat gained by copper = 20 × 377 × 430
= 3 242 200 J or 3·2422 MJ

2. In order to normalise a steel component having a mass of 5 kg it is cooled from 900 °C to 15 °C. If the specific heat of steel is 460 J/kg °C calculate the heat lost by the component.

Drop in temperature = $900° - 15° = 885$ °C

Heat lost by component = $5 \times 460 \times 885$
$$= 2\ 035\ 500 \text{ J or}$$
$$2 \cdot 0355 \text{ MJ}$$

Heat exchanges. In all heat exchange problems the following equation is used
heat lost by hot body = heat gained by cold body.

Example. In order to harden a steel component having a mass of 2 kg it was quenched in water from a temperature of 800 °C. The water had an initial temperature of 20 °C and had a mass of 80 kg. Determine the final temperature of the water and the component. (Specific heat of steel is 460 J/kg °C and of water is 4187 J/kg °C.)

Let final temperature of water and component = t °C.
Then temperature drop of component = $800° - t°$
and heat lost by component = $2 \times 460 \times (800 - t)$
$$= 736\ 000 - 920\ t$$
Temperature rise of water = $t° - 20°$
Heat gained by water = $80 \times 4187 \times (t - 20)$
$$= 334\ 960\ t - 6\ 699\ 200$$
Heat gained by water = heat lost by component
$$334\ 960\ t - 6\ 699\ 200 = 736\ 000 - 920\ t$$
$$334\ 960\ t + 920\ t = 736\ 000 + 6\ 699\ 200$$
$$335\ 880\ t = 7\ 435\ 200$$
$$t = \frac{7\ 435\ 200}{335\ 880} = 22 \cdot 17$$

Hence the final temperature of the water is 22·17 °C.

Heat exchanges and cutting fluids. One of the reasons for using a cutting fluid is to carry away some of the heat generated by the cutting operation. If it is assumed that all the heat energy at the tool point is carried away by the cutting fluid then
heat generated at the toolpoint per second (joules) = heat carried away by the cutting fluid per second (joules).
Heat carried away by cutting fluid per second (joules) = rate of flow (kilogram per second) × specific heat of fluid (joules per kilogram °C) × temperature rise of fluid (°C).

Examples

1. A cutting fluid is flowing at a rate of 5 kg/s. If its specific heat is 4000 J/kg °C and it sustains a temperature rise of 3 °C find the heat carried away by the fluid per second.

Heat carried away by the fluid = $5 \times 4000 \times 3$
$$= 60\ 000 \text{ J/s or } 60 \text{ kW}$$

2. During a turning operation 300 Nm of mechanical energy are converted into heat energy at the tool point every second. If the flow of coolant is 3 kg/min find its temperature rise. (Specific heat of coolant = 3800 J/kg °C.)

Heat energy at toolpoint = 300 Nm/s = 300 J/s.
If temperature rise of coolant = t °C

then heat carried away by coolant = $\dfrac{3}{60} \times 3800 \times t$
$$= 190\ t \text{ joules per second}$$
$$\therefore 190\ t = 300$$
$$t = \frac{300}{190} = 1 \cdot 57$$

Hence the temperature rise of the coolant is 1·57 °C.

Exercise 2.4

1. An aluminium bar having a mass of 20 kg is heated from 20 °C to 200 °C. How much heat has the bar absorbed? (Specific heat of aluminium is 920 J/kg °C)

2. In order to harden a steel component having a mass of 15 kg it is cooled from 820 °C to 20 °C. Calculate the amount of heat lost by the steel. (Specific heat of steel is 460 J/kg °C)

3. A brass forging having a mass of 8 kg is heated from 20 °C to 500 °C. Find the heat gained by the brass if its specific heat is 380 J/kg °C.

4. In order to harden a steel component it is heated and plunged into a tank containing 8 kg of water whose temperature is 15 °C. If the steel loses 800 kJ of heat energy during cooling find the final temperature of the water. (Specific heat of water is 4200 J/kg °C.)

5. A steel component whose mass is 2·5 kg is hardened by heating it to a temperature of 800 °C and quenching it in water. If the water tank contains 80 kg of water at 15 °C calculate the final temperature of the water after the quenching operation. (Specific heats of steel and water are 460 and 4200 J/kg °C respectively.)

6. A steel part having a mass of 2 kg and specific heat 460 J/kg °C is to be heated to 750 °C and plunged into oil of specific heat 1400 J/kg °C and temperature 20 °C. Find the mass of oil required if its temperature rise is not to exceed 5 °C.

7. A cutting fluid is flowing at a rate of 2 kg/s. If its specific heat is 3600 J/kg °C calculate the power absorbed by the fluid if its temperature rises by 2 °C.

8. The power absorbed by a cutting fluid is 4 kW. If its specific heat is 4000 J/kg °C and it sustains a temperature rise of 2 °C find the rate of flow of the fluid in kilograms per second.

Effect of heat on fluids. When a fluid is heated its temperature increases until the fluid starts to change into a gas. Whilst the fluid is changing into a gas its temperature remains constant (at the boiling temperature of the fluid) although heat is still being applied.

When water is heated its temperature increases until it boils and starts to change into steam. The temperature then remains constant until all the water has changed into steam. The temperature at which steam is formed is called the *saturation temperature* and its value depends upon the pressure. At atmospheric pressure the saturation temperature is 100 °C but at a pressure of 4 bars the saturation temperature is about 145 °C.

The heat that is applied to increase its temperature is called *sensible heat* and the heat needed to change the water into steam is called *latent heat*. Thus latent heat is the heat given whilst the temperature remains constant. The latent heat of steam varies with pressure, being about 2200 J/kg at atmospheric pressure and about 2130 J/kg at a pressure of 4 bars.

Total heat = sensible heat + latent heat

Example. 20 kg of water at 15 °C is turned into steam at 100 °C. If the specific heat of water is 4187 J/kg °C and the latent heat of steam is 220 kJ/kg find the total heat required.

Sensible heat = mass of water × specific heat × temp. rise
= 20 × 4187 × 85
= 711 790 J or 711·79 kJ (say 712 kJ)
Latent heat = latent heat per kilogram × mass in kg
= 2200 × 20 = 44 000 kJ
Total heat = latent heat + sensible heat
= 44 000 + 712 = 44 712 kJ

Wet steam. It sometimes happens that insufficient heat is applied to turn all of the water into steam in which case the steam is described as wet steam. To indicate the state of the steam we use the *dryness fraction*. Thus if the dryness fraction is 0·9 in 1 kilogram we would have 0·9 kg of steam and 0·1 kg of water.

Total heat of wet steam = sensible heat + latent heat × dryness fraction

Example. What quantity of heat is required to produce 2 kg of wet steam whose dryness fraction is 0·8 from water at 20 °C? Take the latent heat of steam as 2100 kJ/kg and the saturation temperature as 150 °C.

Sensible heat = mass × specific heat × temperature rise
= 2 × 4187 × 130
= 1 089 000 J or 1089 kJ
Latent heat = mass × latent heat per kilogramme × dryness fraction
= 2 × 2100 × 0·8 = 3360 kJ
Total heat = sensible heat + latent heat
= 1089 + 3360 = 4449 kJ

Latent heat of fusion. When a solid melts and changes into a liquid the heat required to change the solid into a liquid is called the latent heat of fusion. Whilst the solid is changing into a liquid its temperature remains at the melting point of the solid although heat is still being applied.

Example. A billet of aluminium alloy has a mass of 20 kg. Calculate the quantity of heat required to completely melt the billet if its original temperature is 20 °C and the melting point is 660 °C. (Specific heat of the alloy is 0·88 kJ/kg °C and the latent heat of melting is 400 kJ/kg.)

Sensible heat = 20 × 0·88 × (660 − 20)
= 20 × 0·88 × 640 = 11 264 kJ
Latent heat = 20 × 400 = 8000 kJ
Total heat = sensible heat + latent heat
= 11 264 + 8000 = 19 264 kJ

Exercise 2.5

1. 10 kg of water at 15 °C is turned into steam at 100 °C. Find the total heat required if the specific heat of water is 4187 J/kg °C and the latent heat of steam is 2200 kJ/kg.
2. What quantity of heat is required to produce 5 kg of dry steam from water at 20 °C? Take the latent heat of steam as 2140 kJ/kg, the saturation temperature as 120 °C and the specific heat of water as 4187 J/kg °C.
3. Find the quantity of heat required to produce 4 kg of wet steam whose dryness fraction is 0·9 from water at 15 °C. Take the latent heat of steam as 2200 kJ/kg, the saturation temperature as 100 °C and the specific heat of water as 4187 J/kg °C.
4. How much heat is required to produce 5 kg of steam which is 0·85 dry from feed water at 15 °C? The saturation temperature is 165 °C and the latent heat of steam is 2100 kJ/kg. Take the specific heat of water as 4187 J/kg °C.
5. The specific heat of copper is 377 J/kg °C and its latent heat of fusion is 200 kJ/kg. Find the total heat required to completely melt 8 kg of copper if its original temperature is 20 °C and the melting temperature is 1980 °C.
6. An aluminium alloy billet having a mass of 15 kg is completely melted. If its original temperature was 20 °C and the melting point of the alloy is 660 °C find the quantity of heat applied. (Specific heat and latent heat of fusion of the alloy are 880 J/kg °C and 400 kJ/kg respectively.)

Expansion of gases. The volume of a gas can be altered by changing its temperature or by changing its pressure, or by changing its pressure and temperature at the same time. The relationship between volume, pressure and temperature is

$$\frac{p_1 V_1}{T_1} = \frac{p_2 V_2}{T_2}$$

where p_1 = initial pressure
p_2 = final pressure
V_1 = initial volume
V_2 = final volume
T_1 = initial temperature (in kelvins)
T_2 = final temperature (in kelvins)

Note that the units of pressure and volume are unimportant provided the *same* units are used throughout the calculation.

Absolute and gauge pressure. The reading given by most types of pressure gauge is not the actual pressure of the gas but is the difference between the pressure of the gas and the pressure of the atmosphere. The gauge reading is usually called the gauge pressure whilst the actual pressure of the gas is called the absolute pressure. In all calculations it is the absolute pressure which must be used and

absolute pressure = gauge pressure+atmospheric pressure

It will be near enough to take atmospheric pressure as 1 bar or 100 000 N/m².

Examples

1. 2000 cm³ of gas at 9 bars gauge and 27°C expands to 4000 cm³ when the gauge pressure is 7 bars. Calculate the temperature of the gas after the expansion.

   We have p_1 = 9 bars gauge = 10 bars absolute
   $$p_2 = 7 \text{ bars gauge} = 8 \text{ bars absolute}$$
   $$V_1 = 2000 \text{ cm}^3$$
   $$V_2 = 4000 \text{ cm}^3$$
   $$T_1 = 27°C = 273+27 = 300 \text{ K}$$

   Since $\dfrac{p_1 V_1}{T_1} = \dfrac{p_2 V_2}{T_2}$

   $$\frac{10 \times 2000}{300} = \frac{8 \times 4000}{T_2}$$

   $$T_2 = \frac{8 \times 4000 \times 300}{10 \times 2000} = 480$$

Hence the final temperature is 480 K or 480−273 = 207°C

2. A quantity of gas originally occupies 9000 cm³ at 2800 kN/m² absolute pressure and 127°C. It is heated and allowed to expand until its volume and temperature are 18 000 cm³ and 227°C. Find the new absolute pressure.

   We have, p_1 = 2800 kN/m², V_1 = 9000 cm³ and T_1 = 400 K. Also V_2 = 18 000 cm³ and T_2 = 500 K.

   $$\frac{p_1 V_1}{T_1} = \frac{p_2 V_2}{T_2}$$

   $$\frac{2800 \times 9000}{400} = \frac{p_2 \times 18\,000}{500}$$

   $$p_2 = \frac{2800 \times 9000 \times 500}{400 \times 18\,000}$$

   $$= 1750$$

Hence the new pressure is 1750 kN/m² absolute. (Note that the unit of pressure for p_2 must be the same as the unit of pressure used for p_1.)

3. A cylinder contains 0·20 m³ of oxygen at a pressure of 3 MN/m² absolute. If the pressure is reduced to 1·5 MN/m² without change of temperature find the new volume.

   We have p_1 = 3 MN/m², p_2 = 1·5 MN/m² and V_1 = 0·20 m³.

   Since the temperature does not change $T_1 = T_2$ and the relationship

   $$\frac{p_1 V_1}{T_1} = \frac{p_2 V_2}{T_2}$$

   becomes $p_1 V_1 = p_2 V_2$
   $$3 \times 0·20 = 1·5 V_2$$
   $$V_2 = \frac{3 \times 0·20}{1·5} = 0·40 \text{ m}^3$$

Exercise 2.6

1. 5 m³ of gas at 4 bars gauge pressure and 20°C expands to 6 m³ when the pressure is 3 bars gauge. Calculate the temperature of the gas after the expansion.

2. 3 m³ of gas at 6 bars absolute pressure and 30°C expands to 5 m³ when the temperature is 200°C. Calculate the pressure after the expansion.

3. A cylinder of gas contains 2000 cm³ when the pressure is 8 bars gauge. When the pressure is increased to 12 bars gauge what is the volume of gas in the cylinder?

4. A cylinder contains 0·10 m³ of gas at a pressure of 100 kN/m² absolute. At what pressure will the volume be 0·15 m³?

5. The volume of gas in an engine cylinder is 0·1 m³ at 1 bar pressure and 17°C temperature. What will the volume become when the temperature and pressure are increased to 7 bars and 67°C. (The pressures are absolute.)

6. An oxygen cylinder contains 3 m³ at 2·7 MN/m². Find its volume (a) when the pressure decreases to 1·4 MN/m² (b) when the pressure is atmospheric (i.e. 100 kN/m²). All pressures are gauge.

7. A gas cylinder on being raised in temperature from 15°C to 20°C gives an increase in pressure of 0·6 bars. What was the original pressure in the cylinder?

8. A compressed air container has a capacity of 30 m³. It contains air at a temperature of 30°C and a pressure of 8 bars absolute. Find the volume that the air will occupy at 20°C and 2 bars absolute.

2.3 Forces

Equilibrant. The body shown in Fig. 2.4 has three forces applied to it. A single force of $40+30-50 = 20$ N applied to the point X and acting to the right will have exactly the same effect on the body as the three forces originally applied. This force of 20 N is called the *resultant force*.

If the body is free to move then the effect of the three forces is to move the body in the direction of the resultant force, i.e. to the right. The only way to arrest the body is to apply a force of 20 N to the point X acting to the left. Thus by applying a force equal in magnitude to the resultant force—but opposite in direction—the body is placed in a state of *equilibrium*. The force required to produce this state of equilibrium is called the *equilibrant*. Consequently the equilibrant is equal in magnitude to the resultant but it acts in the opposite direction.

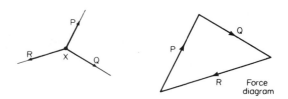

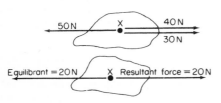

Fig. 2.4 The equilibrant is equal in magnitude but opposite in direction to the resultant force.

The triangle of forces. When three forces acting on a body are in equilibrium they either meet at a point or they are parallel to each other. The triangle of forces deals with three forces which meet at a point and which are in equilibrium. The theorem states that *if a point is in equilibrium under the action of three forces these three forces when represented in a force diagram form a closed triangle*. Thus in Fig. 2.5 the three forces P, Q and R when represented in a force diagram form a closed triangle and hence the point X is in equilibrium. Note that in drawing the force diagram the arrows denoting the senses of the forces must follow head to tail.

Fig. 2.5 The three forces P, Q and R when represented in a force diagram form a closed triangle. Hence the point X is in equilibrium.

The examples which follow show how the triangle of forces is used.

Example. Find the equilibrant of the two forces shown in Fig. 2.6.

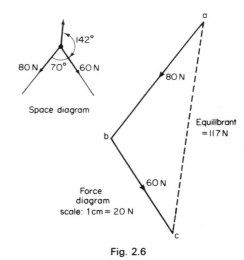

Fig. 2.6

The procedure is as follows:
(1) Draw the space diagram. Only the angles are drawn accurately but the magnitude and direction of each of the forces should be placed on the diagram.
(2) Choose a suitable scale to represent the magnitude of the forces. In this example a scale of 1 cm = 10 N has been chosen.
(3) Using set-squares and a rule draw a line parallel to

the 80 N force 8 cm long and place on it an arrow indicating the sense of the force. Since the arrows must follow nose to tail, from point b draw a line parallel to the 60 N force and 6 cm long. Finally join the points c and a. The length of this line is 11·7 cm and hence the magnitude of the equilibrant is 117 N and its sense is as shown by the arrow. To find its direction draw the dotted line on the space diagram parallel to ac. Hence by using a protractor the direction of the equilibrant is found to act at 142° to the 60 N force.

Example. Fig. 2.7 shows the forces acting on a lathe tool when it it is facing the end of a bar. Find the resultant force acting on the tool.

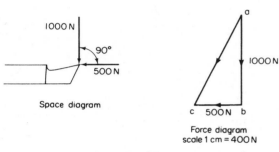

Fig. 2.7

The triangle of force is constructed in the way shown previously. ab represents the 1000 N force and bc the 500 N force. With the arrow pointing from c to a the line ac represents the equilibrant hence with the arrow pointing from a to c the line ac represents the resultant force. By scaling the diagram this is found to be 1120 N acting at 64° to the horizontal.

Example. A load whose weight is 5000 N is supported by two chains as shown in Fig. 2.8. Find the force in each chain.

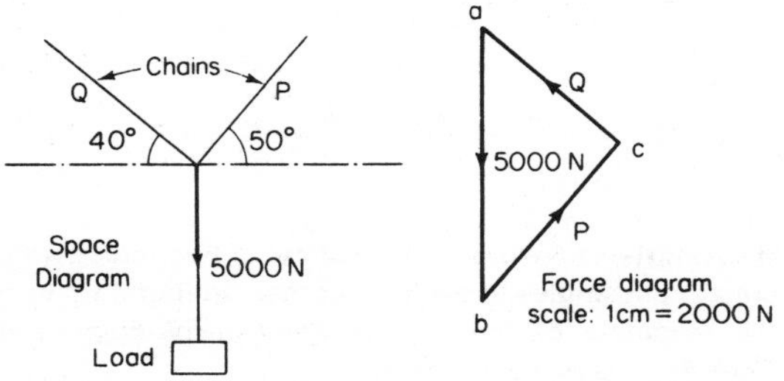

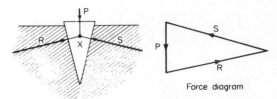

Fig. 2.8

To draw the force diagram first draw ab to represent the 5000 N force. From b draw a line parallel to chain *P* and from a draw a line parallel to chain *Q*. These two lines meet at c and abc is the required triangle of forces. By scaling the diagram the forces in *P* and *Q* are found to be 3800 N and 3300 N respectively.

The wedge. Fig. 2.9 shows a wedge with a vertical force *P* forcing it into position. The forces *R* and *S* which act at right-angles to the faces of the wedge are easily found by using the triangle of forces. Note that since the forces *P*, *R* and *S* are in equilibrium they must meet at a point (point X in the diagram).

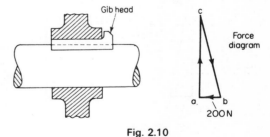

Fig. 2.9 The triangle of forces used to find the forces acting on the faces of a wedge.

Example. A pulley is keyed to a shaft by means of a gib head key which has a taper of 1 in 100 (Fig. 2.10). If a force of 200 N is used to fit the key find the vertical force on the bottom of the keyway.

Fig. 2.10

In the force diagram ac represents the vertical force on the bottom of the keyway whilst bc represents the force on the boss of the pulley and is at right-angles to the top of the key. Because of the small angle of taper it is not practical to scale the force diagram and it is better to calculate the length of ac. The angle acb is the same as the angle of taper and hence

$$\frac{ac}{ab} = \frac{100}{1}$$

Since ab represents a force of 200 N

$$\frac{ac}{200} = \frac{100}{1}$$
$$ac = 200 \times 100 = 20\,000 \text{ N}$$

Hence the vertical force on the bottom of the keyway is 20 000 N.

The wedge may also be used for clamping as shown in the next example.

Example. Fig. 2.11 shows a wedge clamp operated by an air cylinder. Determine the clamping force when the force supplied by the air cylinder is 800 N.

As the wedge is moved to the left by the force from the air cylinder the top wedge slides upwards in the guides thus firmly fixing the component between the top wedge and the jig plate. The force that causes clamping is the

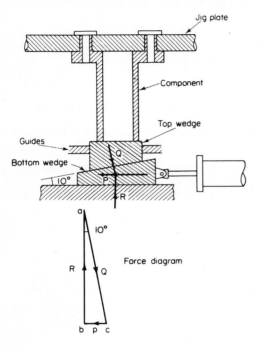

Fig. 2.11

force *R* whilst the force *P* tends to thrust the top wedge against the guides. To release the component the air jack is reversed thus pulling the wedge to the right.

By sketching the force diagram it can be seen that

$$\frac{ab}{bc} = \cot 10°$$

or $\dfrac{R}{P} = \cot 10°$

$$R = P \cot 10° = 800 \times 5 \cdot 67 = 4540 \text{ N}$$

Thus the wedge provides an efficient clamping device in that a comparatively small effort produces a large clamping force.

Clamping. The forces acting on hinged clamps may be found by using the triangle of forces.

Example. Fig. 2.12 shows a hinged clamp. A force *F* applied as shown produces a clamping force *W*. If the force *F* is 600 N find the clamping force *W* and the force *R* acting on the hinge.

The first step is to draw the space diagram to scale. Since the clamp is in equilibrium the forces *F W* and *R* must meet at a point (point X in the diagram). In order to find the magnitudes of *W* and *R* the triangle of forces is drawn as shown and by scaling *W* = 400 N and *R* = 720 N.

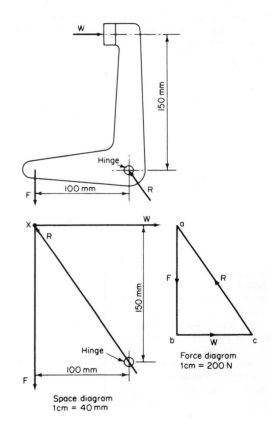

Fig. 2.12

Rectangle of forces. The resultant of two forces which are at right-angles to each other may be found by using the rectangle of forces (see *Mechanical Engineering Craft Studies part 2* (page 40)).

Example. Find the resultant of the two forces shown in Fig. 2.13(a).

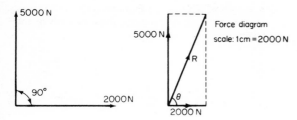

Fig. 2.13

The force diagram is shown in Fig. 2.13(b). By scaling the diagram we find that the resultant force R is 5390 N and it makes an angle of 68° with the horizontal.

Rectangular components of a force. Two forces which meet at a point can be compounded into a single force (the resultant force) which will have the same effect as the two forces. Hence a single force can be split up into two component forces which will have the same effect as the single force.

Thus, in Fig. 2.14 the single force P can be split up into two component forces by using the rectangle of forces. The two component forces are V, the vertical component and H, the horizontal component. The component forces may be calculated as follows:

$$V = P \sin \theta \quad \text{and} \quad H = P \cos \theta$$

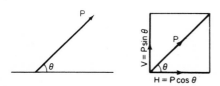

Fig. 2.14   The vertical and horizontal components of a force

Example. Fig. 2.15(a) shows the force applied to a cold chisel by a hammer blow. Find the forces V and H.

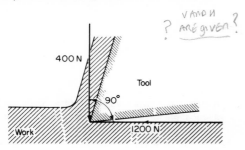

Fig. 2.15a

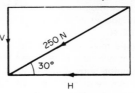

Fig. 2.15b

The forces V and H, the rectangular components of the 250 N force, may be found by drawing a force diagram (Fig. 2.15(b)) or by calculation as follows:

$$V = 250 \times \sin 30° = 250 \times 0.5000 = 125 \text{ N}$$
$$H = 250 \times \cos 30° = 250 \times 0.8660 = 216.5 \text{ N}$$

Example. Fig. 2.16 shows a toggle mechanism which is used in a clamping arrangement. When the force F is 5000 N find the clamping force W and the force R on the slides.

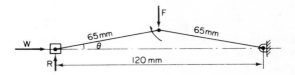

Fig. 2.16

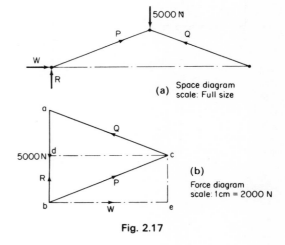

Fig. 2.17

The first step is to draw the space diagram (Fig. 2.17(a)). Next the triangle of forces abc is drawn from which the forces P and Q in the linkages can be found (Fig. 2.17(b)). The final step is to find the vertical and horizontal components of P by using the rectangle of forces becd. The horizontal component of P is W and by scaling the force diagram this is found to be 6000 N. The vertical component of P is R and by scaling this is found to be 2500 N.

With the toggle mechanism, as the links approach a straight line a considerable force is exerted at the slider. For the configuration shown in Fig. 2.16 when $\theta = 10°$, $W = 14\,200$ N and when $\theta = 5°$, $W = 28\,400$ N and when θ approaches $0°$ the force W becomes very large indeed. Thus the advantage of the toggle mechanism is that a very large force is obtained at the slider for a comparatively small force at the handle.

Exercise 2.7

1. Find the equilibrant for each of the systems of forces shown in Fig. 2.18.
2. Fig. 2.19 shows the forces acting on the supports of a travelling steady. Find the resultant force.

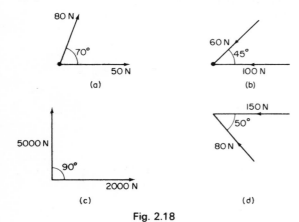

(a)   (b)

(c)   (d)

Fig. 2.18

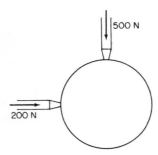

Fig. 2.19

3. A weight of 8000 N is supported on two chains as shown in Fig. 2.20. Find the force in each of the chains.

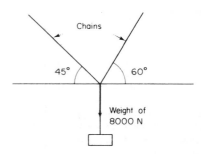

Fig. 2.20

4. A pulley is fixed to a shaft by means of a key (Fig. 2.21) which has a taper of 1 in 80. If a force of 250 N is used to fit the key find the vertical force on the bottom of the keyway.

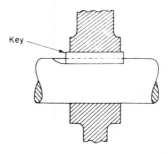

Fig. 2.21

5. In chipping across a surface a fitter holds his chisel at an angle of 40° to the horizontal. If the force acting along the axis of the chisel is 300 N find the horizontal and vertical components of this force.
6. Fig. 2.22 shows the forces acting on a parting-off tool. Find the resultant of the two forces.

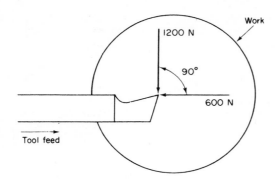

Fig. 2.22

7. Fig. 2.23 shows a billet whose weight is 15 000 N being lifted by a sling. Find the forces in the parts of the sling marked P and Q.

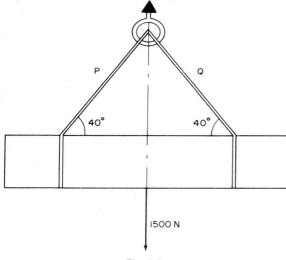

Fig. 2.23

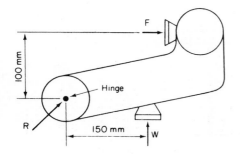

Fig. 2.24

8. Fig. 2.24 shows a hinged clamp in which a force F cause a clamping force W to be produced. Find the force W and the force R acting on the hinge when F is 500 N.

9. For the clamp shown in Fig. 2.25 find the clamping force W and the hinge force R when the force F applied to the clamp is 200 N.

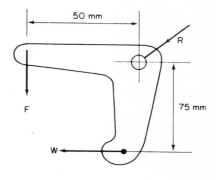

Fig. 2.25

10. In a slider crank mechanism (see Fig. 2.26) the force in the connecting rod is 900 N when the connecting

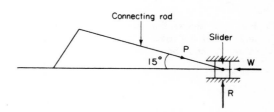

Fig. 2.26

rod makes an angle of 15° with the line of action of the slider. Find the force moving the slider along its guides and the reaction at the guides (i.e. W and R).

11. Fig. 2.27 shows a quick return mechanism which is used on a shaping machine. If the force in the link A is 600 N find the force moving the ram of the machine along its slideways (i.e. force P).

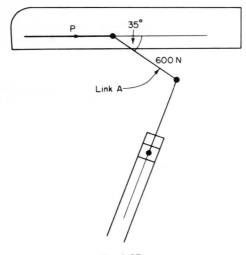

Fig. 2.27

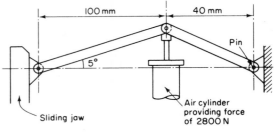

Fig. 2.30

12. In the mechanism shown in Fig. 2.28 find the forces *W* and *R* acting on the slider when the force *P* is 1600 N.

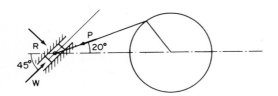

Fig. 2.28

13. Fig. 2.29 shows a toggle mechanism used in a machine. If the force *P* is 2000 N find the forces *W* and *R* acting on the slider.

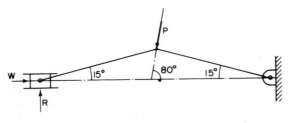

Fig. 2.29

14. Fig. 2.30 shows a pneumatically operated toggle arrangement. Find the force closing the sliding jaw when the links are positioned as shown in the diagram.

15. Fig. 2.31 shows a slotted link used on a shaping machine. Find the force tending to move the roller along the slot when the force in the link is 500 N.

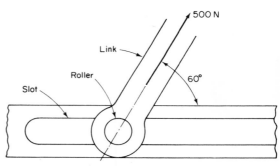

Fig. 2.31

16. Fig. 2.32 shows a screw-driven wedge clamp. Find the force *W* causing the clamping.

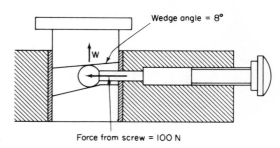

Fig. 2.32

2.4   Moment and torque

Moment. The moment of a force is a measure of its turning effect and its value is calculated from the product of the force and its distance from the fulcrum measured perpendicular to the line of action of the force. Thus in Fig. 2.33 the moment in each case is $F \times d$.

Unit of moment. The unit by which a moment is quoted depends upon the unit of force and the unit of distance as follows:
force in newtons—distance in metres—moment in
newton metres
force in newtons—distance in millimetres—moment in
newton millimetres

Principle of moments. Moments may be either clockwise or anti-clockwise depending upon the direction in which a lever rotates or tends to rotate. Thus in Fig. 2.33 (a) the moment is clockwise and in Fig. 2.33 (b) it is anti-clockwise.

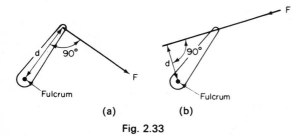

Fig. 2.33

When the sum of the clockwise moments acting on a body exactly equals the sum of the anti-clockwise

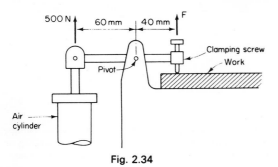

Fig. 2.34

moments then rotation of the body is prevented. This statement is usually called the principle of moments.

Example. Fig. 2.34 shows a clamping system. If the force from the air cylinder is 500 N find the clamping force F.
Taking moments about the pivot,
Clockwise moment = $500 \times 60 = 30\,000$ Nmm
Anti-clockwise moment = $40 \times F$
Since, when the clamp is operating, there must be no rotation
Anti-clockwise moment = clockwise moment
$$40 \times F = 30\,000$$
$$F = 750 \text{ N}$$

Reaction. The beam shown in Fig. 2.35 will not rotate since the clockwise moment about the support equals the anti-clockwise moment about the support. However, the force on the support must exactly equal the downward forces acting on the beam and the upward force which exactly balances the downward forces acting on the beam is called the reaction at the support. Thus in Fig. 2.35 the reaction at the support is $300 + 500 = 800$ N. Note that the actual force at the support will be opposite

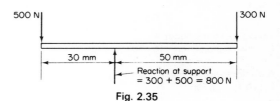

Fig. 2.35

in direction to the reaction.

Example. Fig. 2.36 shows the forces acting on a lever. Calculate the force P to just balance the system and determine the reaction at the pivot.

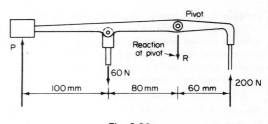

Fig. 2.36

Taking moments about the pivot:

Anti-clockwise moments = $60 \times 80 + 200 \times 60$

$\qquad\qquad\qquad\qquad = 16\,800$ N mm

Clockwise moments = $P \times 180$

For balance:

Clockwise moments = Anti-clockwise moments

$$180\,P = 16\,800$$
$$P = 93\text{ N}$$

To find reaction at pivot:

Total of downward forces = total of upward forces

$$60 + R = 93 + 200$$
$$R = 93 + 200 - 60$$
$$R = 233\text{ N}$$

Beam reactions. Many beams are supported at two points and the problem usually is to determine the reactions at the supports. The solution is always obtained by taking moments about one or other of the support points.

Example. Fig. 2.37 shows the forces acting on the camshaft of an automatic lathe. Find the reactions at the bearings of the shaft.

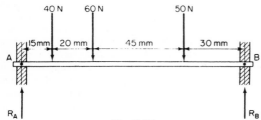

Fig. 2.37

Taking moments about A,

Clockwise moments = $15 \times 40 + 35 \times 60 + 80 \times 50 =$

$\qquad\qquad\qquad\qquad\qquad\qquad$ 6700 Nmm

Anti-clockwise moments = $110\,R_B$

$\therefore 110\,R_B = 6700$

$\qquad R_B = 61$ N

Since total upward forces = total downward forces

$$R_A + R_B = 40 + 60 + 50$$
$$R_A + 61 = 150$$
$$R_A = 89\text{ N}$$

Hence the reactions at the bearings are 61 N and 89 N.

Torque. A body can be made to rotate by applying a turning moment or torque to it. Torque is simply another name for turning moment but it is usual to use the word torque when referring to the turning moment on shafts, spindles, gear wheels etc. In these cases the pivot is the centre of rotation and hence

torque = force × radius

The unit of torque is the same as the unit of turning moment and it usually measured in newton metres (Nm).

Example. During a milling operation there is a torque of 80 Nm acting on the arbor of the machine. If a 120 mm diameter cutter is being used find the force acting on the teeth of the cutter (Fig. 2.38).

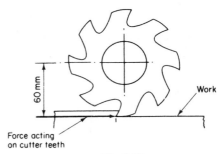

Fig. 2.38

Torque at arbor = Force at cutter teeth × radius of cutter

$$80 = \text{Force} \times \frac{60}{1000}$$
$$\therefore \text{ Force} = \frac{80 \times 1000}{60} = 1333\text{ N}$$

The force acting at the cutter teeth is 1333 N.

Cam clamping. Eccentric clamps make use of cams in order to achieve their clamping action. A force *F* applied to the handle of the clamp (Fig. 2.39) causes a torque at the pivot of the clamp and this gives rise to a large clamping force *W*. The size of the clamping force depends upon several factors but the most important are the distances a and b shown on the diagram. The smaller the values of a and b the greater is the clamping force.

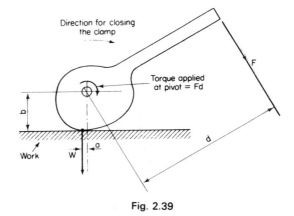

Fig. 2.39

However the large clamping force may cause damage the the workpiece and hence direct cam clamping is only used for unfinished surfaces such as sand castings.

The arrangements shown in Figs. 2.40 and 2.41 are often used instead. Note that in Fig. 2.41 by placing the bolt nearer to the work than to the cam the clamping force on the work is greater than the vertical force supplied by the cam.

Forces in mechanisms. By considering the torque applied to part of a mechanism the force acting on a rack or a ram etc. may be obtained.

Example. Fig. 2.42 shows part of the driving mechanism for a planning machine. If the torque supplied by the motor is 200 Nm find the force *F* driving the rack.

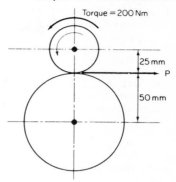

Fig. 2.43

From Fig. 2.43

$$P \times 25 = 200 \times 50$$
$$P = \frac{200 \times 1000}{25} = 8000 \text{ N}$$

$$\text{Torque on worm wheel} = \frac{8000 \times 50}{1000} = 400 \text{ Nm}$$

Since the torque on the pinion is also 400 Nm (Fig. 2.44)

$$F \times 100 = 400 \times 1000$$
$$F = \frac{400 \times 1000}{100} = 4000 \text{ N}$$

Hence the force driving the rack is 4000 N.

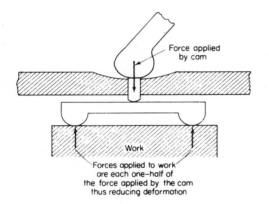

Fig. 2.40

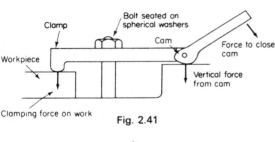

Fig. 2.41

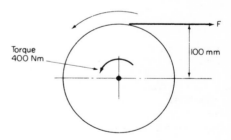

Fig. 2.44

Example. Fig. 2.45 shows the slotted lever of a shaping machine. If the torque applied to the bull wheel is 120 Nm find the force acting on the ram at the mid-point of the stroke (a) on the cutting stroke (b) on the return stroke. Since the torque is 120 Nm, the force on the slider is found from

$$F \times \frac{100}{1000} = 120$$

$$F = 120 \times \frac{1000}{100} = 1200 \text{ N}$$

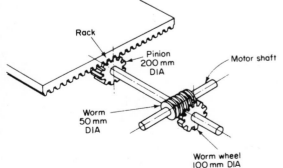

Fig. 2.42

The conditions during the cutting stroke are shown in Fig. 2.46(a). Taking moments about B:

$$600\,R_A = 1200 \times 400$$
$$R_A = \frac{1200 \times 400}{600} = 800\ N$$

Hence the force acting on the ram is 800 N. Fig. 2.46(b) shows the conditions for the return stroke and taking moments about B:

$$600\,R_A = 1200 \times 200$$
$$R_A = \frac{1200 \times 200}{600} = 400\ N$$

Hence, on the return stroke, the force acting on the ram is 400 N.

Compound levers. Compound levers are used extensively in various mechanisms and the method of dealing with them is shown in the following example.

Example. Fig. 2.47 shows a lever system used on a bar automatic machine. A force of 200 N is required as shown for the system to operate. What force F is required?

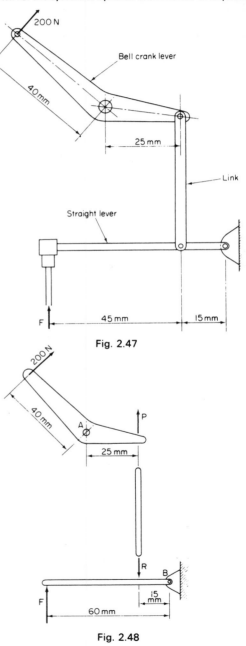

Fig. 2.47

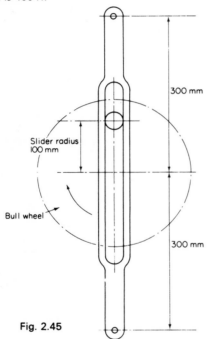

Fig. 2.45

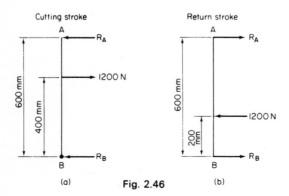

Fig. 2.46

Fig. 2.48

The solution is obtained by drawing each part separately as shown in Fig. 2.48.

First find *P* by considering the bell crank lever. Thus, taking moments about A,

$$P \times 25 = 200 \times 40$$
$$P = \frac{200 \times 40}{25} = 320 \text{ N}$$

This force *P* will be transmitted to the link and at the other end there will be a force equal to *P* but acting in the opposite direction. This is the force *R*.

To find the force *F* we consider the straight lever by itself and by taking moments about B:

$$F \times 60 = R \times 15$$
$$F \times 60 = 320 \times 15$$
$$F = \frac{320 \times 15}{60} = 80 \text{ N}$$

Hence the force *F* required to operate the mechanism is 80 N.

Exercise 2.8

1. Fig. 2.49 shows a bell crank lever which is used on an automatic lathe. A force of 150 N is required as shown for the lever to operate. What force *F* is required on the other arm?

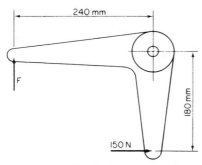

Fig. 2.49

2. Fig. 2.50 shows a screw-type clamping device. Find the force clamping the component. What is the force at the hinge?

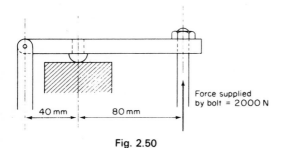

Fig. 2.50

3. Fig. 2.51 shows the forces acting on a lever. Calculate the force *F* required to balance the system and find the reaction at the pivot.

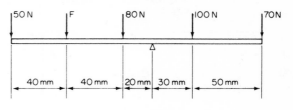

Fig. 2.51

4. Fig. 2.52 shows a tool post. Calculate the forces *R₁* and *R₂* at the screws when the cutting force at the tool point is 1500 N.

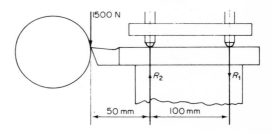

Fig. 2.52

5. Fig. 2.53 shows a shaft supported in bearings. The mass of the shaft is 100 kg and it may be assumed to be concentrated at the point shown in the diagram. Calculate the reaction at each bearing.

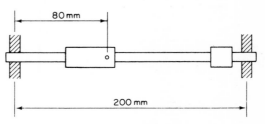

Fig. 2.53

6. A bar is being turned by using the front and rear toolposts simultaneously. Fig. 2.54 shows the cutting forces involved. Calculate the vertical reactions at each of the centres.

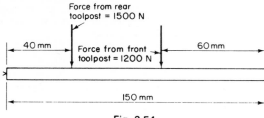

Fig. 2.54

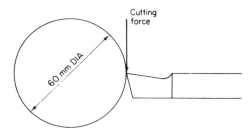

Fig. 2.57

7. Fig. 2.55 shows diagrammatically the forces involved when cutting a piece of material held in a chuck. Calculate the reactions at the spindle bearings.

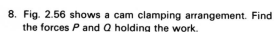

Fig. 2.55

8. Fig. 2.56 shows a cam clamping arrangement. Find the forces *P* and *Q* holding the work.

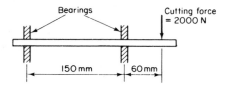

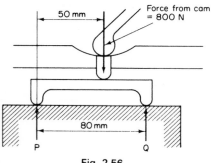

Fig. 2.56

9. The motor of a centre lathe supplies a torque of 120 Nm at its spindle. If work of 60 mm diameter is being turned calculate the maximum cutting force that can be used (Fig. 2.57).

10. Fig. 2.58 shows a rack and pinion mechanism. If the torque on the pinion is 150 Nm calculate the force *F* on the rack.

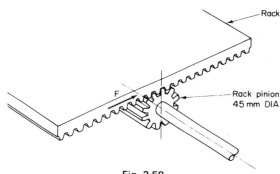

Fig. 2.58

11. Fig. 2.59 shows the gear mechanism used when hand traversing a lathe carriage. If the tangential force applied at the handwheel is 40 N calculate the force *F* on the rack. Assume that there are no losses due to friction etc.

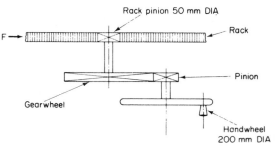

Fig. 2.59

12. Fig. 2.60 shows a lifting machine. Find the force required tangentially at the handle in order to lift a load of 500 N. Assume no losses due to friction etc.

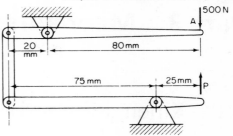

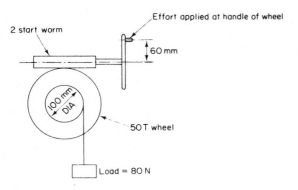

Fig. 2.60

13. Fig. 2.61 shows a system of levers. Calculate the force *P* required in order to just overcome the force of 500 N at point A.

14. The diagram (Fig. 2.62) shows details of a pneumatically operated clamping system. If the force supplied by the air cylinder is 600 N find the clamping force *F*.

Fig. 2.61

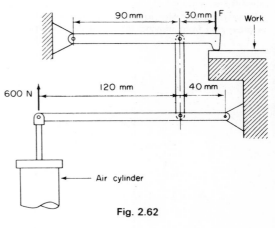

Fig. 2.62

Part 3   Mechanisms

3.1   Kinematics and alignment

Kinematic principles. As shown in Fig. 3.1 there are six possible ways in which a body can move. These are called the six degrees of freedom.

By utilizing six restrictions suitably applied a body may be prevented from moving in any of the six directions.

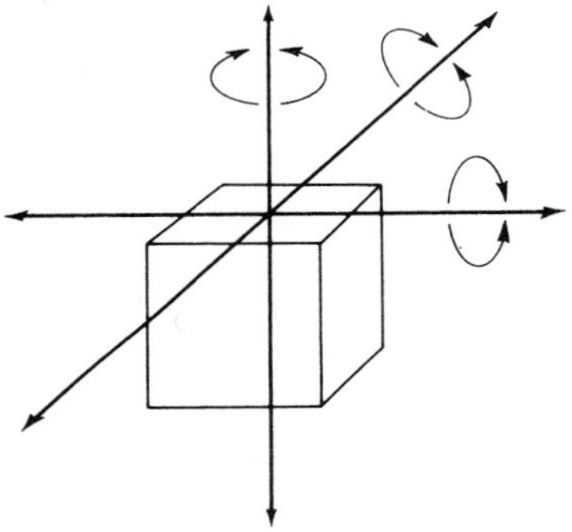

Consider the shaft in Fig. 3.2 which passes through the hole in the bracket. It is constrained from moving in all six directions. However, if the stop is removed (i.e. one of the restrictions is removed) then the shaft can slide lengthways. When the key is also removed (i.e. two of the restrictions are removed) the shaft can rotate as well as moving lengthways.

Sliding pairs. When the stop in Fig. 3.2 is removed the shaft can slide because of the key which is a sliding fit in the keyway of the bracket. Sliding can also be achieved by having a rectangular element moving in a rectangular hole in the bracket (Fig. 3.3) whilst in Fig. 3.4 the slotted bar slides over the die block. Each of these are known as sliding pairs and it is seen that in each case movement is possible in one direction only.

Fig. 3.1   The six degrees of freedom.

Removing one of these restrictions will permit movement in one direction, removing two will permit movement in two directions etc.

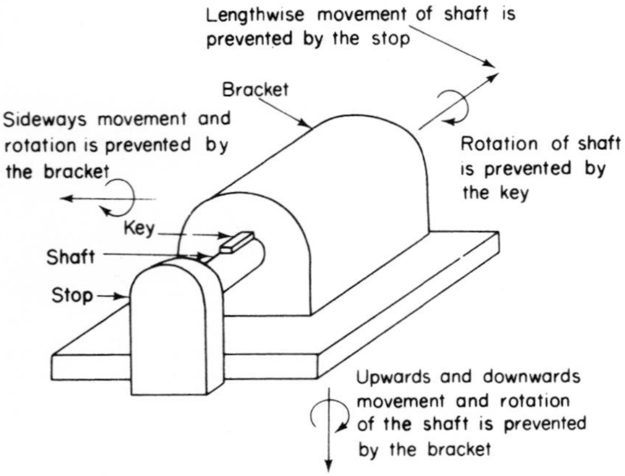

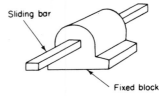

Fig. 3.3

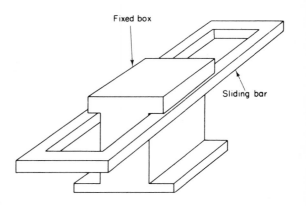

Fig. 3.2   The shaft is constrained from moving in all six directions.

Fig. 3.4

It is important that only the minimum number of constraints is used. For instance in Fig. 3.5 it is not necessary for the slider to fit over the two vee-slideways. One vee is sufficient to prevent movement in a sideways direction. It is seen that the weight of the slider is sufficient to prevent movement in an upwards direction. Gravity is often the only force needed but in some cases clamping may be necessary.

The basic alignment tests for machine tools are given in BS 4656 and they are as follows:
1. The checking of tables and slideways.
2. The checking of spindles.
3. The checking of the alignment of the spindle with another feature.
4. Performance tests.

To give an idea of the type of tests used and the methods of carrying out the tests we will consider each of the above as applied to a centre-lathe.

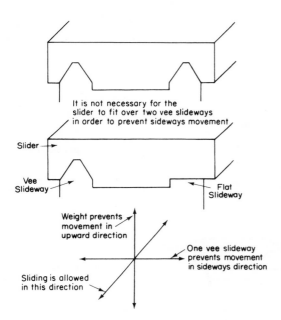

Fig. 3.5 Only the minimum number of constraints should be used. One vee guide is sufficient to prevent sideways movement.

Slideways for machine tools. The slides and slideways locate and guide parts of the machine tool which move relative to each other. They are usually used for altering the position of the cutting tool relative to the workpiece.

Slideways vary according to the purpose for which the machine tool is intended and some typical slides and slideways are shown in Fig. 3.6. Notice that in every case only the minimum number of constraints is used.

Alignments of machine tools. If the slides and slideways are not in the correct relationship to each other then it will be impossible to produce accurate work. Alignment tests are performed for this reason.

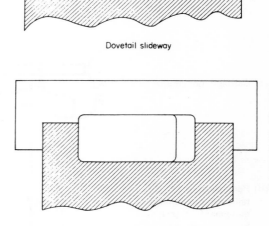

Dovetail slideway

Flat slideway

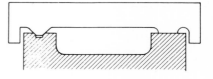

Vee slideway

Fig. 3.6 Types of slideways.

1. TABLES AND SLIDEWAYS

Test	Diagram	Permissible error
(a) Straightness of slideways	(a)	Convex 0·006 mm ⎫ per ⎬ 300 Concave 0·013 mm ⎭ mm
(b) Flatness of slideways	(b)	±0·006 per 300 mm
(c) Carriage movement	(c)	0·006 per 300 mm
(d) Parallelism of tailstock and carriage movements. Check in both vertical and horizontal planes	(d)	0·006 per 300 mm

(Dean, Smith and Grace Ltd.)

2. SPINDLE

Test	Diagram	Permissible error
(a) End float	(a)	0·007 mm
(b) Run-out of spindle nose external taper	(b)	0·010 mm
(c) Run-out of flange face	(c)	0·010 mm
(d) Run-out of centre point	(d)	0·010 mm

(Dean, Smith and Grace Ltd.)

3. ALIGNMENT OF SPINDLE AXIS.

Test	Diagram	Permissible error
(a) Axis of spindle parallel with bed in the vertical plane.		0·02 mm per 300 mm
(b) Axis of spindle parallel with bed in the horizontal plane.		0·02 mm per 300 mm
(c) Axis of tailstock and headstock centres parallel with the bed in the vertical plane.		0·02 mm per 300 mm

(Dean, Smith and Grace Ltd.)

4. PERFORMANCE TESTS

The recommended performance tests for a centre lathe are as follows:

1. Test for round turning using a chuck.
2. Test for parallel turning using a chuck. Allowable tolerance is 0·03 mm per 300 mm.
3. Test for parallel turning between centres. Allowable tolerance is 0·02 mm in any length.
4. Test for facing. The allowable tolerance is 0·02 mm per 300 mm diameter.

Errors caused by lack of alignment. Some of the effects of errors in alignment are given below for two types of machine tools.

1. The lathe. The important things are that the machine should be capable of producing turned and bored work which is straight and parallel and facing work which is flat.

Alignment error	Effect on the work
Axis of spindle not parallel with bed in the horizontal plane.	Work will be turned with a taper.
Axis of spindle not parallel with bed in the vertical plane	Work will be turned with a concave surface and it may also be tapered.
Cross-slide does not move perpendicularly to axis of spindle.	Work will be faced with either a convex or a concave surface.

2. The horizontal milling machine. The machine should produce faces which are parallel with the machine table and the cutters should cut to size.

Alignment error	Effect on the work
Internal taper of spindle runs out of true.	Cut is not shared equally between the teeth of the cutter.
End play in spindle	Work will not be machined to size because the cutters float sideways.
Transverse movement of table not parallel with spindle axis.	Milled face will not be parallel with the base of the work.
Centre T-slot is not square with spindle axis and not parallel with table movement.	Work is located by means of the T-slot. Hence work will not have faces in correct relationship with each other.
Table not perpendicular to vertical slideways.	Vertical faces on work will not be at right-angles to the base of the work.
Arbor not straight.	Cutters will not run true.

Exercise 3.1

1. Fig. 3.7 shows several different kinds of slideways. For each state how movement in the remaining five degrees of freedom is prevented.

2. State the effect that the following misalignments (on a drilling machine) would have
   (a) Column not square with the base in the plane of the drill.
   (b) Drill head guides not square with table.
   (c) Spindle sleeve not square with table.

3. The following questions relate to a grinding machine.
   (a) How would you check the vertical slide for squareness?
   (b) How would you check the straightness of the slideways?

4. By means of a diagram show how you would check the alignment of the spindle of a vertical milling machine with the table.

5. State the principal alignment tests that are necessary for a centre-lathe. Show how two of these tests are made.

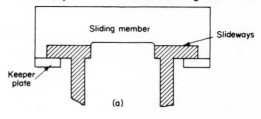

(a)

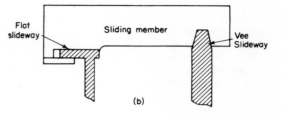

(b)

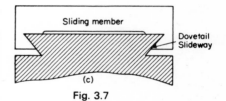

(c)

Fig. 3.7

3.2 Mechanisms

The control of machine tools. The important controls are as follows:

1. Control of the spindle speed. This usually achieved by means of pulleys and gears.
2. Control of slides. This may be achieved in a variety of ways, for example by mechanical or hydraulic means.

Belt drives. Fig. 3.8 shows a compound belt drive.

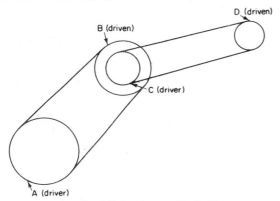

Fig. 3.8 A compound belt drive

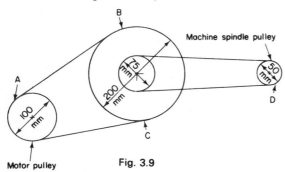

Fig. 3.9

The speed ratio between the first and last pulleys may be found by using the relationship:

$$\frac{\text{speed of driver pulley}}{\text{speed of driven pulley}} = \frac{\text{diameters of driven pulleys}}{\text{diameters of driver pulleys}}$$

Thus for the drive of Fig. 3.8.

$$\frac{\text{speed of A}}{\text{speed of D}} = \frac{\text{diameter of B} \times \text{diameter of D}}{\text{diameter of A} \times \text{diameter of C}}$$

Example. Fig. 3.9 shows the drive between an electric motor and the spindle of a machine tool. If the motor pulley revolves at 800 rev/min find the spindle speed of the machine tool.

$$\frac{\text{speed of A}}{\text{speed of D}} = \frac{\text{diameter of B} \times \text{diameter of D}}{\text{diameter of A} \times \text{diameter of C}}$$

$$\frac{800}{\text{speed of D}} = \frac{200 \times 50}{100 \times 75}$$

$$\frac{800}{\text{speed of D}} = \frac{4}{3}$$

$$\text{speed of D} = 800 \times \frac{3}{4} = 600$$

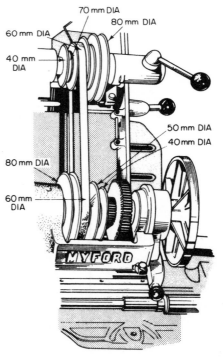

Fig. 3.10   *(Myford Ltd.)*

Hence the spindle speed of the machine tool is 600 rev/min.

A range of spindle speeds may be obtained by using stepped pulleys as shown in the following example.

Example. Fig. 3.10 shows part of the drive for a small lathe. If the top shaft has a speed of 240 rev/min find the four speeds of the spindle shaft which can be obtained.

$$\text{1st speed } = 240 \times \frac{40}{80} = 120 \text{ rev/min}$$

$$\text{2nd speed } = 240 \times \frac{60}{60} = 240 \text{ rev/min}$$

$$\text{3rd speed } = 240 \times \frac{70}{50} = 336 \text{ rev/min}$$

$$\text{4th speed } = 240 \times \frac{80}{40} = 480 \text{ rev/min}$$

A variable speed drive may be obtained by using two cones as shown in Fig. 3.11. To reduce the speed of

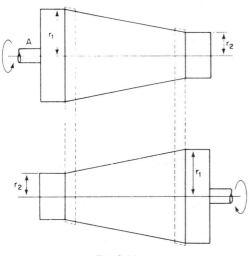

Fig. 3.11

shaft B the belt is moved to the right along the cones. This type of drive overcomes the limitations of the fixed speed ratio which is obtained with the normal belt drive. Thus in Fig. 3.11 the ratio $\dfrac{\text{speed of shaft A}}{\text{speed of shaft B}}$ is infinitely variable between $\dfrac{r_1}{r_2}$

Example. In a variable speed drive (Fig. 3.11) the speed of shaft A is 200 rev/min. If $r_1 = 60$ mm and $r_2 = 30$ mm, find the range of speeds which can be obtained.

The highest speed that can be obtained is given by:

$$\frac{\text{speed of shaft B}}{\text{speed of shaft A}} = \frac{r_1}{r_2}$$

$$\frac{\text{speed of shaft B}}{200} = \frac{60}{30}$$

$$\text{speed of shaft B} = \frac{60}{30} \times 200 = 400 \text{ rev/min}$$

$$\frac{200}{\text{speed of B}} = \frac{2}{1}$$

$$\text{speed of B} = \frac{200}{2} = 100 \text{ rev/min}$$

The lowest speed that can be obtained is given by:

$$\frac{\text{speed of B}}{\text{speed of A}} = \frac{r_2}{r_1}$$

$$\frac{\text{speed of B}}{200} = \frac{30}{60}$$

$$\text{speed of B} = \frac{1}{2} \times 200$$

$$\text{speed of B} = 100 \text{ rev/min}$$

Thus the speed of shaft B is infinitely variable between 100 rev/min and 400 rev/min.

Gear drives. Most gear drives consist of a train of gears (Fig. 3.12). The speed ratio between the first and last wheels is found by using the formula:

$$\frac{\text{speed of driver}}{\text{speed of driven}} = \frac{\text{number of teeth in driven wheels}}{\text{number of teeth in driver wheels}}$$

Thus, for the train in Fig. 3.12,

$$\frac{\text{speed of A}}{\text{speed of D}} = \frac{\text{number of teeth in B} \times \text{number of teeth in D}}{\text{number of teeth in A} \times \text{number of teeth in C}}$$

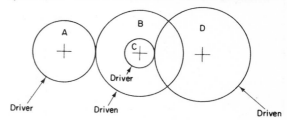

Fig. 3.12   A train of gears

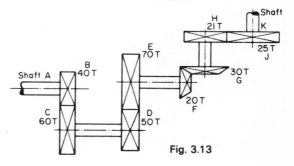

Fig. 3.13

Example. Fig. 3.13 shows the gear drive for a machine tool. If shaft A rotates at 900 rev/min find the speed of shaft K.

$$\frac{\text{speed of A}}{\text{speed of K}} = \frac{T_C \times T_E \times T_G \times T_J}{T_B \times T_D \times T_F \times T_H}$$

where T_C = number of teeth in wheel C
T_E = number of teeth in wheel E etc.

$$\frac{900}{\text{speed of J}} = \frac{60 \times 70 \times 30 \times 25}{40 \times 50 \times 20 \times 21}$$

$$\frac{900}{\text{speed of J}} = \frac{15}{4}$$

$$\text{speed of J} = \frac{900 \times 4}{15} = 240 \text{ rev/min}$$

Spindle speed changes may be obtained by using sliding gears on a splined shaft (Fig. 3.14). The disadvantage of this method is that the spindle must be synchronised or stopped before a speed change is made. Rapid gear changes without stopping the spindle are possible when plate clutches (Fig. 3.15) are used because the gears are always in mesh.

Example. In Fig. 3.16 find all the speeds available from the gear box given:

Gear A	35 T	B	50 T	C	25 T
	D	65 T	E	50 T	F	75 T
	G	55 T	H	70 T	K	45 T

Speed number	Gear arrangement	Spindle speed
1.	A 35 T / D 65 T / G 55 T	$\frac{35}{55} \times 900 = 573$ rev/min
2.	A 35 T / D 65 T — E 50 T / H 70 T	$\frac{35 \times 50}{65 \times 70} \times 900 = 346$ rev/min
3.	A 35 T / D 65 T — F 75 T / K 45 T	$\frac{35 \times 75}{65 \times 45} \times 900 = 808$ rev/min
4.	B 50 T / E 50 T / H 70 T	$\frac{50}{70} \times 900 = 643$ rev/min
5.	B 50 T / E 50 T — D 65 T / G 55 T	$\frac{50 \times 65}{50 \times 55} \times 900 = 1064$ rev/min
6.	B 50 T / E 50 T — F 75 T / K 45 T	$\frac{50 \times 75}{50 \times 45} \times 900 = 1500$ rev/min
7.	C 25 T / F 75 T / K 45 T	$\frac{25}{45} \times 900 = 500$ rev/min
8.	C 25 T / F 75 T — D 65 T / G 55 T	$\frac{25 \times 65}{75 \times 55} \times 900 = 354$ rev/min
9.	C 25 T / F 75 T — E 50 T / H 70 T	$\frac{25 \times 50}{75 \times 70} \times 900 = 214$ rev/min

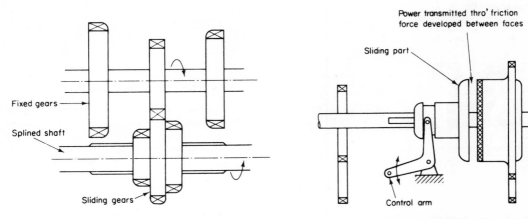

Fig. 3.14 Sliding gears on a splined shaft

Fig. 3.15 Plate clutches give rapid gear changes without stopping the spindle

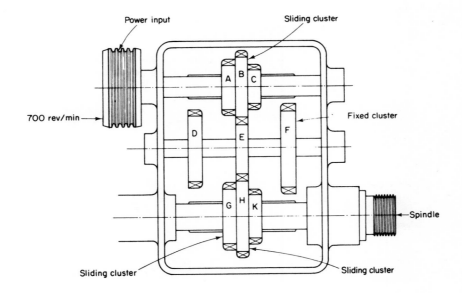

Fig. 3.16

In the feed gearboxes a range of speeds in close ratio is required. This is frequently achieved by using a Norton gearbox (Fig. 3.17).

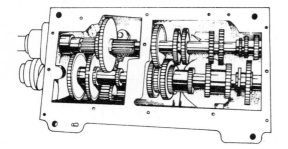

(Kerry Machine Tools Ltd.)
Fig. 3.17   Feed gear-box

Mechanisms used for controlling slides and tables. The problem in controlling the motion of slides and tables is that of converting rotary motion into linear motion. Some of the methods used are:
1. The rack and pinion (Fig. 3.18).
2. The leadscrew and nut.
3. The cam (Fig. 3.19).
4. Linkages (Fig. 3.20).

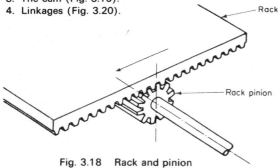

Fig. 3.18   Rack and pinion

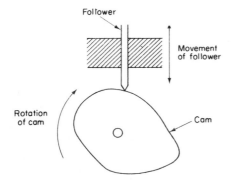

Fig. 3.19   Cam and follower

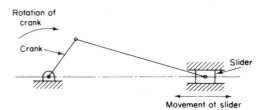

Fig. 3.20   Slider crank mechanism

Rack and pinion. Each tooth of the pinion meshes in turn with one tooth on the rack. Hence in one revolution of the pinion:

distance moved by rack = number of teeth on the pinion
× pitch of the rack teeth.

Example. In Fig. 3.18 the pinion has 25 teeth and revolves at a speed of 10 rev/min. If the rack teeth have a pitch of 5 mm calculate the speed of the rack.
Distance moved by the rack when the pinion makes 1 revolution

= number of teeth on the pinion × pitch of the rack teeth
= 25 × 5 = 125 mm

Since the pinion has a speed of 10 rev/min.
Distance moved by the rack in 1 min = 125 × 10 = 1250 mm
Hence the speed of the rack = 1250 mm/min or 1·25 m/min.

Frequently a worm and worm wheel (Fig. 3.21) are introduced into a geared system in order to reduce the

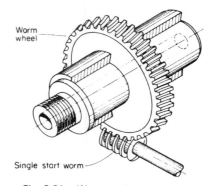

Fig. 3.21   Worm and worm wheel

speed of movement. When the worm is single start one revolution of the worm causes the wheel to move through one tooth. If the worm is multi-start then one revolution of the worm causes the wheel to move through a number of teeth equal to the number of starts on the worm.

Example. Fig. 3.22 shows the mechanism inside a lathe apron which is used to control the feed of the carriage. If shaft A revolves at 32 rev/min calculate the feed of the carriage in millimetres per minute.

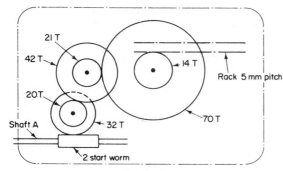

Fig. 3.22

One revolution of the worm causes the 32 T wheel to move through 2 teeth, that is, $\frac{1}{16}$th of a revolution. Hence the speed of the 32 T wheel is $\frac{1}{16} \times 32 = 2$ rev/min. Since the speed of the 20 T wheel is the same as that for the 32 T wheel

$$\frac{\text{speed of 20 T wheel}}{\text{speed of 70 T wheel}} = \frac{42 \times 70}{20 \times 21}$$

$$\frac{2}{\text{speed of 70 T wheel}} = 7$$

speed of 70 T wheel $= \frac{2}{7}$ rev/min

Since the speed of the 14 T wheel is the same as the speed of the 70 T wheel:

Number of teeth engaging with the rack $= \frac{2}{7} \times 14 = 4$ teeth/min

Hence the carriage feed $= 4 \times 5 = 20$ mm/min

Leadscrew and nut. The accurate relationship between the rotation of a screw and its axial movement is frequently used to give a precise movement to the tables and slides of machine tools. For a single start screw the distance moved by the screw in one revolution is equal to the pitch of the screw. If the screw is multi-start the distance moved by the screw in one revolution is equal to the number of starts on the screw.

Example. The mechanism for the feed drive on a plano-milling machine is shown in Fig. 3.23. Calculate the feed in millimetres per minute if the motor makes 480 rev/min.

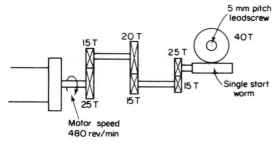

Fig. 3.23

Consider the gear train up to the 25 T wheel:

$$\frac{\text{speed of motor}}{\text{speed of 25 T wheel}} = \frac{15}{25} \times \frac{15}{20} \times \frac{25}{15}$$

$$\frac{480}{\text{speed of 25 T wheel}} = \frac{3}{4}$$

speed of 25 T wheel $= \frac{4}{3} \times 480 = 640$ rev/min

Hence the worm makes 640 rev/min and the speed of the screw is $\frac{640}{40} = 16$ rev/min. Therefore the feed is $16 \times 5 = 80$ mm/min.

Screw cutting on a lathe. On some lathes screw cutting is done with the aid of change wheels (Fig. 3.24).

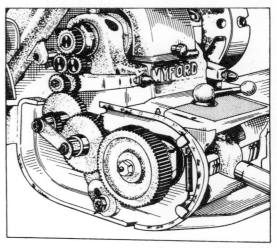

(Myford Ltd.)

Fig. 3.24   Change wheels for thread cutting

The gear wheels required for cutting a specific thread may be calculated from the equation

$$\frac{\textit{teeth in driver wheels}}{\textit{teeth in driven wheels}} = \frac{\textit{pitch of thread to be cut}}{\textit{pitch of leadscrew}}$$

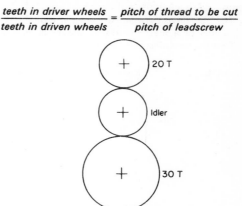

Fig. 3.25

The standard change wheels range from 20 teeth to 120 teeth in 5 T intervals (i.e. 20 T, 25 T, 30 T etc.) One of the smaller wheels is usually duplicated and this is often the 40 T wheel.

Example. Determine suitable change wheels for cutting the following threads: (a) 4 mm pitch (b) 56 mm pitch. The pitch of the leadscrew is 6 mm.

(a) $\dfrac{\text{Teeth in driver wheels}}{\text{Teeth in driven wheels}} = \dfrac{4}{6} = \dfrac{2}{3} = \dfrac{2\times10}{3\times10} = \dfrac{20}{30}$

Hence a simple gear train (Fig. 3.25) is suitable.

(b) $\dfrac{\text{Teeth in driver wheels}}{\text{Teeth in driven wheels}} = \dfrac{56}{6} = \dfrac{28}{3} = \dfrac{7\times4}{3\times1}$

$$= \frac{(7\times10)\times(4\times20)}{(3\times10)\times(1\times20)}$$

$$= \frac{70\times80}{30\times20}$$

Thus to cut a thread with a pitch of 56 mm a compound train (Fig. 3.26) is required.

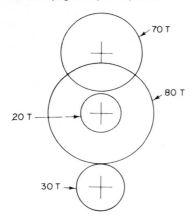

Fig. 3.26

Exercise 3.2

1. The belt drive between the electric motor and the spindle of a machine tool is shown in Fig. 3.27. Find the spindle speed of the machine.

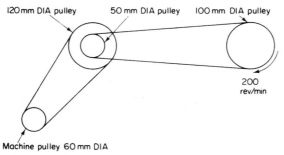

Fig. 3.27

2. Fig. 3.28 shows a compound belt drive. Find the speed of the 100 mm diameter pulley.

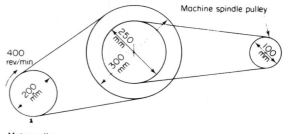

Fig. 3.28

3. Fig. 3.29 shows the drive for a small drilling machine. If the motor speed is 1200 rev/min calculate the four spindle speeds that can be obtained.

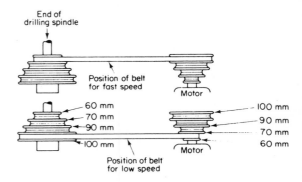

Fig. 3.29

4. In a variable speed drive similar to Fig. 3.11 the speed of shaft A is 500 rev/min. If $r_1 = 50$ mm and $r_2 = 25$ mm, find the range of speeds that can be obtained.

5. Fig. 3.30 shows the drive for a machine tool. If shaft A revolves at 500 rev/min calculate the speed of shaft B.

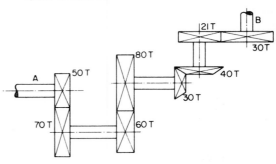

Fig. 3.30

6. In Fig. 3.31 find the speed of shaft B.

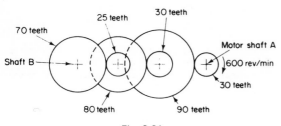

Fig. 3.31

7. In Fig. 3.32 find all the available speeds, given

A 45 T	B 60 T	C 35 T
D 65 T	E 50 T	F 75 T
G 55 T	H 70 T	K 45 T

8. In Fig. 3.33 calculate the distance moved by the rack when the handwheel makes one revolution.

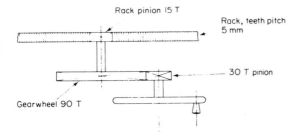

Fig. 3.33

9. Fig. 3.34 shows part of a machine tool drive. Find the table speed when the motor shaft makes 600 rev/min.

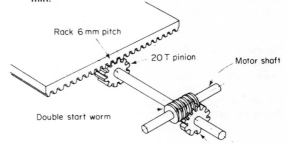

Fig. 3.34

10. Fig. 3.35 shows part of the feed mechanism for a machine tool. Calculate the feed when the motor shaft makes 800 rev/min.

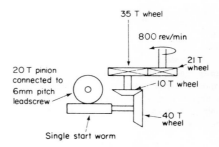

Fig. 3.35

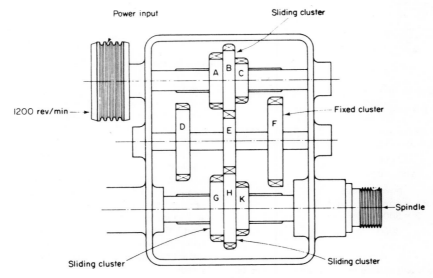

Fig. 3.32

11. Using the standard change wheels find suitable gear trains for cutting threads with the following pitches: (a) 5 mm (b) 8 mm (c) 7·5 mm (d) 12·5 mm (e) 50 mm. The lathe has a leadscrew with a pitch of 6 mm.

Power in mechanisms. It was shown in Chapter 2.1 that for linear motion:

Power (watts) = force (newtons) × speed (metres per second)

and for rotating mechanisms:

Power (watts) = 2π × torque (newton metres) × rotational speed (revolutions per second)

Examples.

1. An electric motor is rated at 5 kW when revolving at 1400 rev/min. Find the torque available at this rating.

Power = 2π × torque × rotational speed

$$5000 = 2 \times \frac{22}{7} \times torque \times \frac{1400}{60}$$

$$5000 = \frac{440}{3} \times torque$$

$$torque = \frac{5000 \times 3}{440} = 34 \text{ Nm}$$

The torque available is 34 Nm

2. The mechanism shown in Fig. 3.36 is driven by a 8 kW motor. If shaft A revolves at 400 rev/min and assuming no power losses find the force *F* acting on the rack.

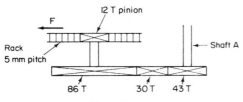

Fig. 3.36

The first step is to find the speed of the rack. Thus

speed of 86 T wheel = $\frac{43}{86} \times 400 = 200$ rev/min.

This will also be the speed of the 12 T pinion. Hence speed of rack = 200 × 12 × 5 = 12 000 mm/min

$$= \frac{12\,000}{1000 \times 60} = 0.2 \text{ m/s}$$

Since there are no power losses

Power at rack = power at motor

$$F \times 0.2 = 8000$$

$$F = \frac{8000}{0.2} = 40\,000 \text{ N or 40 kN}$$

Efficiency. There will always be some power losses in mechanisms because of friction, backlash etc. and

hence no machine can be 100% efficient. The efficiency of a machine is obtained from

$$efficiency \% = \frac{power \ output \ of \ machine}{power \ input \ to \ machine} \times 100$$

Example. Fig. 3.37 shows the drive from the motor to the spindle of a lathe. If the motor develops 5 kW at 700 rev/min find the torque available for cutting if the efficiency of the drive is 75%.

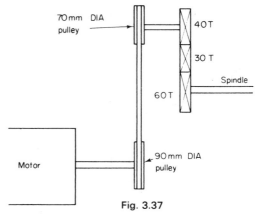

Fig. 3.37

The first step is to calculate the spindle speed. Thus

$$spindle \ speed = 700 \times \frac{90}{70} \times \frac{40}{60} = 600 \text{ rev/min}$$

Power available at spindle = 75% of power at motor
= 0·75 × 5 = 3·75 kW

Let *T* be the torque available at the spindle, then

$$2\pi \times T \times \frac{600}{60} = 3750$$

$$T = \frac{3750}{2\pi \times 10} = 59.7 \text{ Nm}$$

Hence the torque available for cutting is 59·7 Nm.

Exercise 3.3

1. A motor is rated at 3 kW when its speed is 700 rev/min. Calculate the torque available.

2. The tensions on the tight and slack sides of a belt drive are 500 N and 300 N respectively. The pulley is 150 mm diameter and it revolves at 500 rev/min. Calculate the power which the drive transmits.

3. A small electric motor runs at 1400 rev/min. What torque must it exert to develop 2 kW? By means of gearing the motor drives a lathe spindle rotating at 140 rev/min. What torque does the lathe spindle exert? If the work diameter is 100 mm what is the maximum tangential cutting force that can be applied assuming that all the 2 kW is available at the lathe spindle?

4. Fig. 3.38 shows a train of gears which is used in a

lathe headstock. If the motor runs at 1500 rev/min calculate the speed of the spindle.

The motor develops 5 kW and the efficiency of the drive is 75%. Calculate the torque available at the spindle.

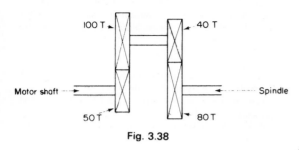

Fig. 3.38

5. Fig. 3.39 shows the feed mechanism for a certain machine tool. Calculate the feed when the shaft A makes 72 rev/min.

The torque available at the shaft A is 50 Nm and the efficiency of the drive is 80%. Calculate the force exerted by the leadscrew.

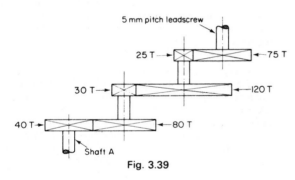

Fig. 3.39

6. Fig. 3.40 shows a rack and pinion mechanism. The torque at the shaft A is 4 Nm and it rotates at 1400 rev/min. If the efficiency is 75% determine the force F exerted by the rack.

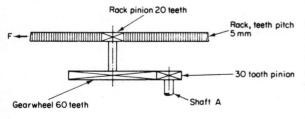

Fig. 3.40

Cams. Cams are used to convert a uniform motion into a determined irregular motion. Usually a cam has a constant angular speed and a typical plate cam is shown in Fig. 3.41 in which the line of action of the follower

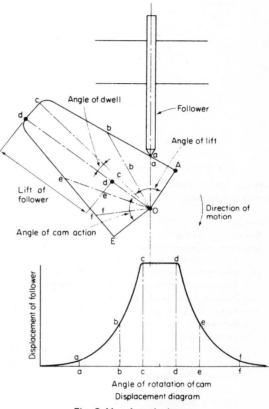

Fig. 3.41 A typical cam

passes through the centre of the cam. When the cam is moving in a clockwise direction the follower is lifting when the cam moves through the angle AOc and it is at rest (or dwelling) when the cam moves through the angle cOd. The follower is falling when the cam moves through the angle dOE.

In order to find the displacement of the follower at any instant a displacement diagram is drawn. With the follower in the position shown the displacement of the follower is aa. Displacements for other positions of the follower are found by measuring radial distances bb, cc etc. The displacement diagram is then drawn by plotting these radial distances against the angle of rotation of the cam.

The wear on a cam may be reduced by using a roller follower but this has the effect of modifying the profile of the cam because the point of contact between the cam and roller is not always on the stroke of the follower (Fig. 3.42).

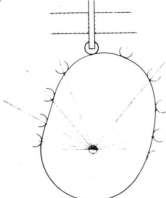

Fig. 3.42   Cam with a roller follower

Setting out cam profiles. The method is shown in the following examples.

Examples.
1. Set out the profile of a cam to give a pointed follower the displacements shown in Fig. 3.43(a). The cam is to have a minimum diameter of 100 mm.
   (1) Mark off the displacement diagram (Fig. 3.43(a)) so that the distances aa, bb . . . mm are obtainable.
   (2) Draw a circle of 100 mm diameter (Fig. 3.43(b)). Mark off the angular lines shown.
   (3) On the 80° line mark off the radial distance aa, on the 100° line the distance bb and so on until on the 340° line the distance mm is marked off.
   (4) Join all the points so obtained. The resulting shape is the required cam profile.

2. Set out the profile of cam to give a roller follower of diameter 20 mm the displacements shown in Fig. 3.44(a). The cam is to have a minimum diameter of 80 mm.
   (1) Mark off the displacement diagram so that the heights aa, bb . . . hh are obtainable.
   (2) Draw a circle of diameter 80 mm and mark off the angular lines shown in Fig. 3.44(b).
   (3) Mark off the radial distances aa, bb . . . hh. Draw the arcs of 20 mm diameter to represent the roller.
   (4) The cam outline is obtained by drawing a curve tangential to the small arcs.

Many cams are made to give constant feed followers. This means that the follower must rise and fall through equal distances for equal angles of rotation of the cam. Hence the displacement diagram must be composed of straight lines as shown in Fig. 3.43(a). Whilst the cam is rotating from 60° to 120° the follower rises at a constant speed and from 180° to 360° the follower falls at a constant speed.

Linkages. A linkage which converts circular motion into a linear motion is the crank and connecting rod mechanism (Fig. 3.20).

In order to obtain a displacement diagram for the slider draw the mechanism to scale with the crank in various positions (Fig. 3.45). The displacement diagram is a curve and hence the motion of the slider is not uniform but its speed varies along the length of the stroke (Fig. 3.46).

Quick return mechanisms. Two kinds of quick return motions are used in machine tools:
1. The slotted link mechanism (Fig. 3.47) which is used on shaping machines.
2. The Whitworth quick return mechanism (Fig. 3.48) which is used on slotting machines.

Slotted link mechanism. For this mechanism

$$\frac{time\ for\ cutting\ stroke}{time\ for\ return\ stroke} = \frac{360-\theta}{\theta}$$

The maximum speed of the ram occurs when the slotted link is vertical on both the cutting and return strokes. When the stroke length is altered (by moving the crankpin) the ratio of cutting time to return time is also altered. The ratio becomes smaller when the length of the stroke is reduced and greater when the length of the stroke is increased.

Example. Fig. 3.49 shows the essentials of a shaping machine mechanism. Find the ratio of cutting time to return time when the crankpin radius is (a) 100 mm (b) 50 mm.

Using the constructions shown in Fig. 3.50 the ratios may be calculated by using trigonometry as follows:
(a) In $\triangle ABC$ (Fig. 3.50(a)),

$\angle ABC = 90°$ (angle between a radius and a tangent)

$BC = 100$ mm and $AC = 300$ mm

$$\therefore\ \cos\frac{\theta}{2} = \frac{100}{300} = 0.3333$$

$$\frac{\theta}{2} = 70°\ 30'$$

$$\theta = 2 \times 70°\ 30' = 141°$$

$$\frac{\text{Cutting time}}{\text{Return time}} = \frac{360°-141°}{141°} = \frac{219°}{141°} = 1.55$$

(b) In $\triangle ABC$ (Fig. 3.50(b))

$$\cos\frac{\theta}{2} = \frac{50}{300} = 0.1667$$

$$\frac{\theta}{2} = 80°\ 30'\ \text{(approx)}$$

$$\theta = 161°$$

$$\frac{\text{Cutting time}}{\text{Return time}} = \frac{360°-161°}{161°} = \frac{199°}{161°} = 1.24$$

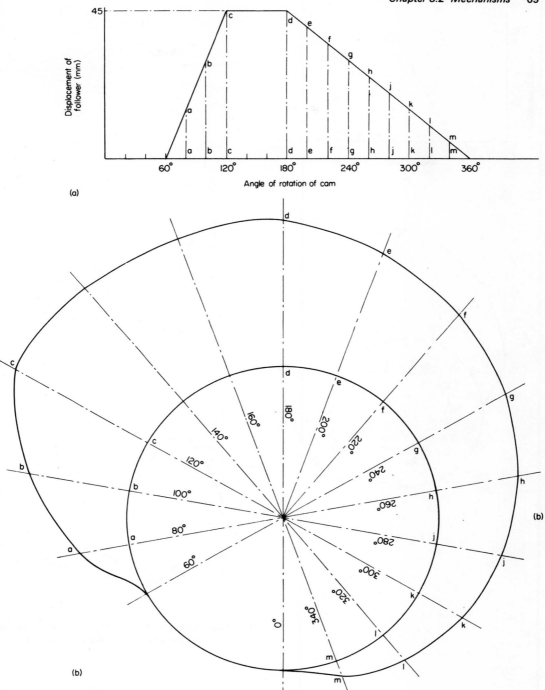

(a)

(b)

(b)

Fig. 3.43

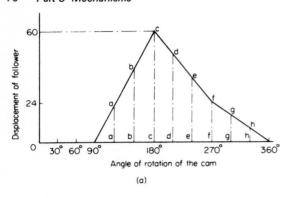

(a)

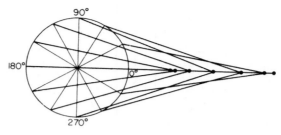

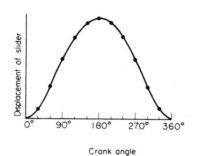

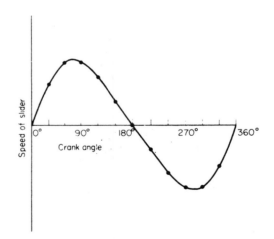

Fig. 3.45   Displacement diagram for the slider of a slider crank mechanism

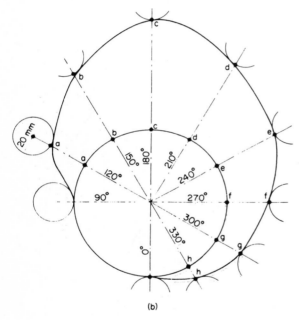

(b)

Fig. 3.44

Fig. 3.46   Diagram showing the variation of the speed of the slider of a slider crank mechanism.

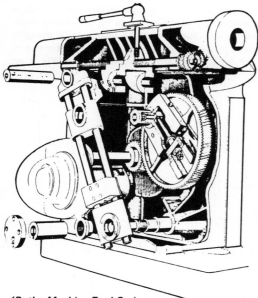

(Butler Machine Tool Co.)

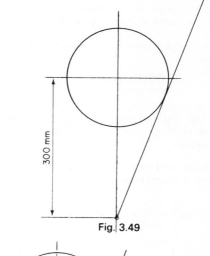

Fig. 3.49

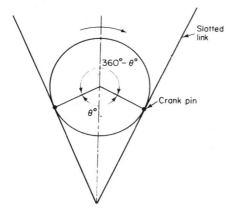

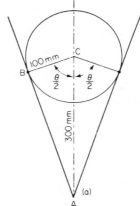

Fig. 3.47 The slotted link mechanism. On the cutting stroke the crank pin moves through the angle 360°−θ° and on the return stroke through the angle θ°.

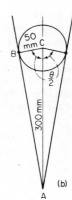

Fig. 3.50

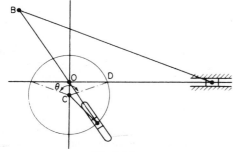

Fig. 3.48

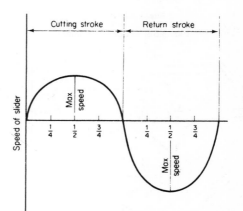

Fig. 3.51 Diagram showing the variation of the speed of the slider of a slotted link mechanism

As with the ordinary slider crank mechanism the velocity of the slider of the slotted link mechanism varies along its stroke as shown in Fig. 3.51.

Whitworth quick return mechanism (Fig. 3.48). For this mechanism:

$$\frac{cutting\ time}{return\ time} = \frac{360° - \theta°}{\theta°}$$

The stroke length is altered by changing the length OB. Since this does not affect the angle θ the ratio of cutting time to return time remains the same for all lengths of stroke.

Example. In Fig. 3.48 if OC = 40 mm and CD = 120 mm find the ratio of cutting time to return time.
In △OCD

$$\cos\frac{\theta}{2} = \frac{OC}{CD} = \frac{40}{120} = 0.3333$$

$$\frac{\theta}{2} = 70° \text{ (approx.)}$$

$$\therefore \theta = 140°$$

$$\frac{\text{Cutting time}}{\text{Return time}} = \frac{360 - \theta}{\theta} = \frac{360 - 140}{140} = \frac{220}{140} = 1.57$$

The four-bar chain. A typical four-bar chain is shown in Fig. 3.52. A practical use for this linkage is the pantograph (Fig. 3.53) which is frequently used on engraving and die-sinking machines.

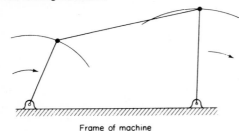

Fig. 3.52   Four bar chain

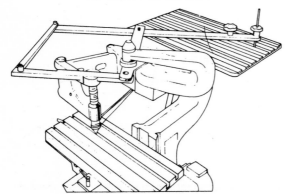

(Alexander Engraving Machines)
Fig. 3.53   Pantograph for an engraving machine

In Fig. 3.54 the tracer of the pantograph will reproduce the reverse shape of the template on a reduced scale. The pantograph ratio is found by using similar triangles as shown in the diagram. The follower, fulcrum and tracer all lie on the same straight line and this is an

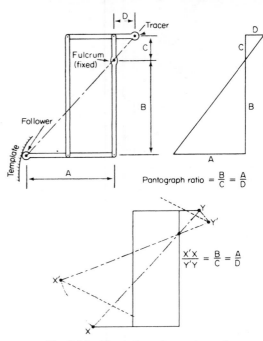

Fig. 3.54   The action of a pantograph

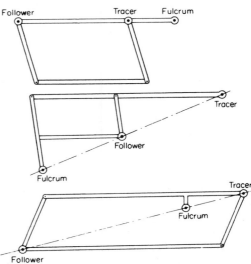

Fig. 3.55   In each of the pantograph systems in the diagram the follower, fulcrum and tracer all lie on the same straight line.

essential feature of any pantograph (see also Fig. 3.55 which shows the types of linkages commonly used in pantograph systems).

Forces acting in linkages. It has been shown in Chapter 2.1 that

$$\text{power} = \text{force} \times \text{speed}$$

If at a point A in a linkage the force is F_A and the speed is v_A then the power at point A is $F_A v_A$. Similarly if at point B the speed is v_B and the force is F_B then the power at point B is $F_B v_B$. If the mechanism is 100% efficient then

$$F_A v_A = F_B v_B$$

otherwise

$$\frac{F_A v_A}{F_B v_B} = \text{efficiency}$$

Example. The torque input at the crank of the mechanism shown in Fig. 3.56 is 80 Nm. Find the force acting on the ram if the crank speed is 1400 rev/min and the ram speed is 120 m/s and the efficiency is 80%.

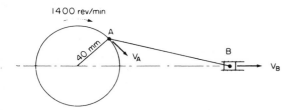

Fig. 3.56

The first step is to calculate the linear speed of the crank pin A. Thus

$$v_A = \frac{2\pi \times 1400}{60} = 147 \text{ m/s}$$

Since

$$\text{torque} = \text{force} \times \text{radius}$$

$$F_A = \frac{80}{0 \cdot 04} = 2000 \text{ N}$$

$$\frac{F_B v_B}{F_A v_A} = \text{efficiency}$$

$$\frac{F_B \times 120}{2000 \times 147} = 0 \cdot 8$$

$$F_B = 2000 \times 0 \cdot 8 \times 147 = 235\ 200 \text{ N or } 235 \cdot 2 \text{ kN}$$

Hence the force acting on the ram is 235·2 kN.

Exercise 3.4

1. Set out the profile of a cam to give a pointed follower the displacements shown in Fig. 3.57. The cam is to have a minimum diameter of 80 mm.

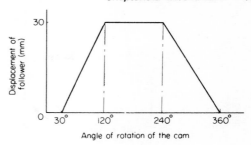

Fig. 3.57

2. A pointed follower is to have the displacements shown in Fig. 3.58. If the cam is to have a minimum diameter of 100 mm set out its profile.

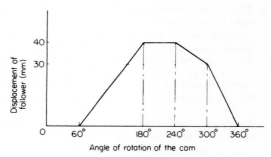

Fig. 3.58

3. Set out the profile of a cam to give a roller follower of diameter 20 mm the displacements shown in Fig. 3.43. The cam is to have a minimum diameter of 80 mm.
4. Draw a constant velocity cam which raises its pointed follower through 40 mm in $\frac{1}{3}$ rev, keeps it there for $\frac{1}{6}$ rev and then lowers it at constant velocity in $\frac{1}{3}$ rev where it rests for the remaining $\frac{1}{6}$ rev. The minimum diameter of the cam is to be 60 mm.
5. Fig. 3.59 shows part of a shaping machine mechanism. Calculate the ratio of cutting time to return time. Find the time required to machine a surface 100 mm wide if the stroke is 300 mm, the feed per stroke is 2 mm and the average cutting speed is 20 m/min.
6. In the quick return motion shown in Fig. 3.48 if OC = 60 mm and CD = 180 mm find the ratio of cutting time to return time. If OB = 240 mm find the stroke of the slider.
7. Draw a displacement diagram for the slider of the mechanism shown in Fig. 3.60.

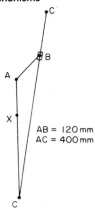

C
B
A
X
AB = 120 mm
AC = 400 mm
C

Fig. 3.59

A
B
O
OA = 60 mm AB = 200 mm

Fig. 3.60

D
E
A
C
B
AB = 30 mm
BC = 80 mm
CD = 300 mm
DE = 200 mm

Fig. 3.61

8. Draw a displacement diagram for the slider of the mechanism shown in Fig. 3.61.
9. Fig. 3.62 shows a pantograph mechanism. Calculate the pantograph ratio. What is the dimension *X*?
10. The torque input at the crank of the mechanism shown in Fig. 3.60 is 100 Nm. Find the force acting on the slider if the crank speed is 1200 rev/min and the slider speed is 100 m/s. Assume the efficiency is 75%.
11. The force acting at the crank pin of a slider-crank mechanism similar to Fig. 3.60 is 3000 N and the linear speed of the pin is 150 m/s. The speed of the slider is 100 m/s and the force acting at the slider is is 2500 N. What is the efficiency of the mechanism?

Follower
120 mm
X
25 mm
Tracer
150 mm

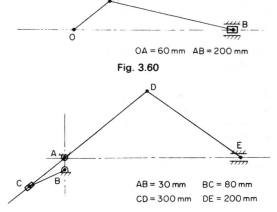

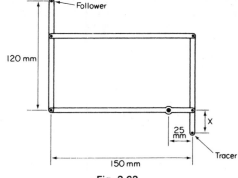

Fig. 3.62

3.3 Vibration

Free vibration. Fig. 3.63 shows a spring with a mass attached to it. If we pull on the mass the spring stretches and when we release the mass suddenly it moves upwards and downwards so that a vibration is caused.

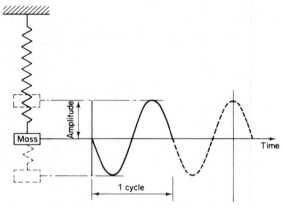

Fig. 3.63   Free vibration of a spring

If the size of the vibration is measured at various intervals of time the wave shown in Fig. 3.63 is obtained. The *amplitude* of the vibration is $\frac{1}{2}$ ×peak to peak distance. The wave form repeats itself and each complete wave constitutes a cycle. The number of cycles that occur in 1 second is called the *frequency* and a typical frequency might be 30 cycles per second. However in the SI system 1 cycle per second = 1 hertz (Hz) and hence the frequency would be 30 Hz. The frequency of the vibration depends upon the size of the mass hung on the spring (the larger the mass the lower the frequency) and the stiffness of the spring (the stiffer the spring the higher the frequency.) Note that the stiffer the spring the less it stretches under a given load.

The vibration of the spring is called a free vibration because once the vibration has been set going it continues without any external force being needed.

Damping. Vibrating systems are all subject to damping. A free vibration, such as is exhibited by the spring in Fig. 3.63, always tends to die down (i.e. the amplitude becomes smaller) because of friction and other resistances, and the vibration is said to be damped (see Fig. 3.64).

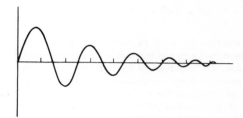

Fig. 3.64   A damped free vibration

Forced vibration. When a vibration takes place under the action of external forces it is called a forced vibration. A forced vibration takes place at the frequency of the force causing vibration.

Natural frequency. If the spring in Fig. 3.63 is considered weightless then the system will always vibrate at the same frequency. This frequency is called the natural frequency of the system.

However, if we have a spring with two masses on it (Fig. 3.65) then we will get two natural frequencies, one for each mass.

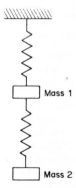

Fig. 3.65   This system has two natural frequencies

Shafts and beams may be regarded as very stiff springs and hence they are capable of being vibrated. The cantilever in Fig. 3.66 if deflected by a force and released will start to vibrate. Similarly, if a torque is applied to a shaft it twists. If the torque is suddenly released the shaft will vibrate torsionally by twisting first one way and then the reverse way.

Resonance test. The cantilever shown in Fig. 3.66 may be regarded as being made up of a great many masses and hence it possesses a great many natural frequencies.

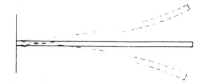

Fig. 3.66   Vibration of a cantilever

These may be found by conducting a resonance test in which the cantilever is vibrated by using a vibrator which allows the frequency of the exciting force to be varied. A point is chosen on the cantilever and for each change of frequency of the exciting force the amplitude at the point is measured. When the amplitude of the point is plotted against the frequency of the exciting force a diagram similar to Fig. 3.67 is obtained where it will be noticed that at certain frequencies the amplitudes peak.

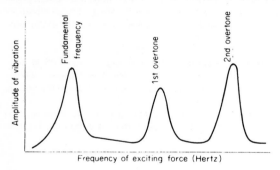

Fig. 3.67   Natural frequencies

The frequencies at these peaks are the natural frequencies of the cantilever and they are sometimes called the fundamental frequency, the first overtone, the second overtone etc.

When the frequency of the exciting force is the same as one of the natural frequencies of the cantilever *resonance* is said to occur and the natural frequencies are therefore called the resonance frequencies.

For each of the natural or resonance frequencies the cantilever will assume a different shape called the mode of vibration. Thus for the cantilever of Fig. 3.66 the modes of vibration are similar to those shown in Fig. 3.68. The natural frequencies depend upon the stiffness of the cantilever in bending and they increase as the stiffness increases.

Machine tool frames. The frame of a machine tool has several natural frequencies, each associated with a definite mode of vibration. For one type of horizontal milling machine the natural frequencies due to horizontal

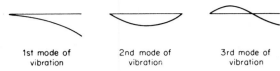

| 1st mode of vibration | 2nd mode of vibration | 3rd mode of vibration |

Fig. 3.68   The modes of vibration of a cantilever

excitation are 20, 45, 55 and 80 Hz. If the exciting force has the same (or nearly the same) frequency as one of the natural frequencies then resonance occurs. The machine alignments may then be seriously affected with the result that poor work will be produced.

Sources of vibration in machine tools. The major sources of vibration are as follows:
1. *The driving motor* which has three sources of vibration: out of balance of the rotating parts, ball bearings and the rotating magnetic field. The amount of vibration can be reduced by isolating the motor from the spindle. This is often achieved by using belt drives.
2. *The gear box* in which the main sources of vibration are misaligned gears and unbalanced gear wheels. The vibrations become more severe when the gears are rotating at high speeds.
3. *Unbalanced face-plates and rotating cutters.* In both cases any out of balance results in vibration. The out of balance forces increase as the speed of rotation increases and in the case of a grinding wheel a small amount of out of balance results in a large exciting force.
4. *Irregular feed motion* can cause vibration particularly when turning.
5. *Irregular cutting.* When milling, each tooth cuts in turn and hence some vibration must occur. This can be minimised by using sharp cutters. Hard spots in castings etc. cause fluctuating forces which can set up vibrations.

Vibration of cutting tools. Chatter marks on a machined surface may be caused by imperfections in the driving gears or the machine bearings but it is frequently caused by vibration of the cutting tool.

When cutting with a discontinuous type of chip there is a periodic variation in the cutting force (Fig. 3.69). When the frequency of the cutting force variations becomes nearly equal to one of the natural frequencies of the tool or the workpiece resonance occurs and severe chatter takes place. In practice large cutting forces, a negative rake angle and a broad thin chip (caused by a small approach angle on a large nose radius) all tend to produce chatter.

Anti-vibration mountings. When a machine is rigidly attached to a supporting structure any vibration from the machine will be transmitted to the support. This often results in undesirable vibrations of the support and

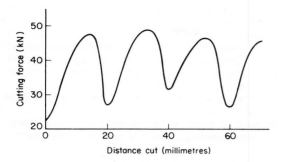

Fig. 3.69   Periodic variation in a cutting force

provision for levelling and alignment. Each mount is capable of supporting about 4000 kg and eliminates all vibration from outside sources.

Whirling of shafts. The centre of gravity of a loaded shaft is always displaced from the axis of rotation owing

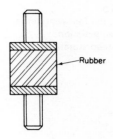

Fig. 3.71   Anti-vibration mounting for light-weight equipment

surrounding structure. In the opposite situation any vibration of a supporting structure will be transmitted to the machine.

In either case, to keep the transmitted vibration to a minimum the source of the vibration must be isolated by means of flexible supports which have been properly designed for the system (Fig. 3.70).

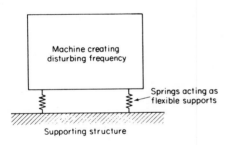

Fig. 3.70   Flexible supports acting as an anti-vibration mounting.

For the most effective vibration isolation the mounting should be such that its natural frequency is low in comparison with that of the disturbance. As a rule the disturbing frequency should always be greater than $2\frac{1}{2}$ times the natural frequency of the mounting. Since most disturbing frequencies are 30 Hz or more, most anti-vibration mountings are designed to produce a natural frequency of 8 to 10 Hz.

In heavy equipment with a flat base several materials are available for vibration isolation. Sheets of felt, cork and rubber are most commonly used and are very effective where very high frequencies are to be reduced. Lightweight motors and instruments can be installed with the type of isolator shown in Fig. 3.71. This isolator has a natural frequency of about 8 Hz. Fig. 3.72 shows an anti-vibration mounting for a machine tool which has

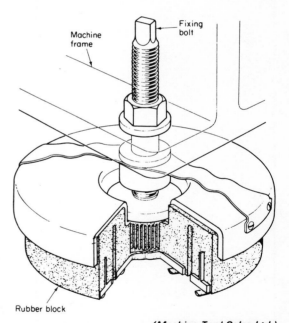

(Machine Tool Sales Ltd.)

Fig. 3.72   Anti-vibration mounting for a machine tool

to lack of straightness of the shaft, bending of the shaft or the eccentric mounting of pulleys, gears etc. with which the shaft is loaded. As the shaft rotates the resulting centrifugal force bends the shaft and hence its centre of gravity is displaced still further. Clearly, the faster the shaft rotates the greater is the effect. However, since the shaft possesses stiffness when it rotates, it tends to vibrate violently in a transverse direction at certain speeds known as whirling speeds or critical speeds. It is only in the neighbourhood of the critical speeds that the effect is noticeable, at speeds between them the shaft rotates normally. The greater the out of balance of the shaft the more violent is the vibration.

Exercise 3.5

1. Explain the following terms:
   (a) Free vibration.
   (b) Forced vibration.
   (c) Damping.
   (d) Natural frequency.
   (e) Resonance.
   (f) State three sources of vibration in a machine tool and suggest methods for reducing the vibration.
2. A lathe tool is subject to severe vibration and chatter. How may this be reduced?
3. Why are anti-vibration mountings used? On what principle do they work?
4. What is meant by the whirling of a shaft? What causes whirling? How may the effect be reduced?

Part 4 Materials

4.1 Strength of materials

Materials under load. Forces may cause a material to extend, to compress, to bend, to shear or to twist. Clearly, the effects that forces (or loads) have on materials are of great importance and these effects are discussed in detail below.

Tensile forces. A tensile force is one which tends to increase the length of the material on which it acts. The material itself is said to be in tension.

Tensile stress. A piece of material which has a force (or load) applied to it is in a state of stress. The intensity of stress (usually just called the stress) is calculated from

$$stress = \frac{applied\ load}{c.s.a.\ of\ material}$$

where c.s.a. stands for cross-sectional area.

If the load is measured in newtons and the cross-sectional area in square millimetres then the stress is measured in newtons per square millimetre (N/mm²). It is, however, usual to state stresses in meganewtons per square metre (MN/m²) and it is worth while remembering that 1 N/mm² = 1 MN/m².

Example. A square bar of steel of 20 mm side is subjected to a load of 10 kN. Calculate the stress in the bar.
Applied load = 10 kN = 10 000 N
  c.s.a = 20×20 = 400 mm²

$$stress = \frac{applied\ load}{c.s.a.} = \frac{10\ 000}{400} = 25\ N/mm²\ or\ 25\ MN/m²$$

Strain. When a material is in a state of stress it becomes deformed. This deformation due to stress is called the strain. For a material in tension the strain is measured by using the equation

$$strain = \frac{amount\ of\ extension}{original\ length\ of\ material}$$

The extension and the original length must be measured in the same units (either millimetres or metres). Strain has no units—it is just a number.

Example. A bar 2 m long has a tensile force applied to it and as a result it extends by 0·24 mm. Calculate the strain.

$$strain = \frac{0·24}{2000} = 0·000\ 12$$

Tensile test. Fig. 4.1 shows a typical testing machine and Fig. 4.2 shows the shapes of standard test pieces for metal. Frequently, during the test, the measurement of very small extensions is necessary and an extensometer (Fig. 4.3) must then be used.

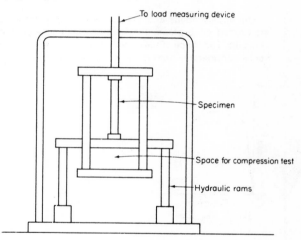

Fig. 4.1 Tensile testing machine

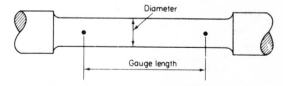

Fig. 4.2 Standard test pieces for the tensile test.

SIZES OF STANDARD TEST PIECES

Diameter (mm)	Area (mm²)	Gauge length (mm)
11·28	100	56
13·82	150	69
15·96	200	80
22·56	400	113

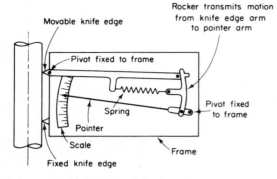

Fig. 4.3   Huggenberger extensometer.

The applied loads and the corresponding extensions are plotted on a graph, the load always being plotted vertically. The shape of the resulting load-extension curve varies considerably from metal to metal (Fig. 4.4).

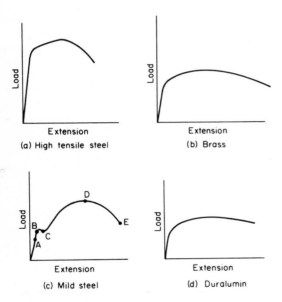

Fig. 4.4   Load-extension curves for various metals.

The important points on the graph for mild steel are:
A. *The limit of proportionality.* Up to this point the graph is a straight-line and hence load is proportional to extension.
B. *The elastic limit.* Up to this point the metal is elastic, i.e. when the load is removed the metal returns to its original length.
C. *The yield point.* For some materials (mild steel is an example) a sudden extension or yielding occurs without any increase in load.

D. *The ultimate load.* This is the maximum load recorded during the test and it corresponds to the ultimate tensile stress which is often called the tensile strength of the material.
E. *The breaking load.* This corresponds to the nominal breaking stress. Although the breaking load is less than the ultimate load the actual breaking stress is greater than the ultimate tensile stress because of the localised necking of the test specimen (see Fig. 4.5).

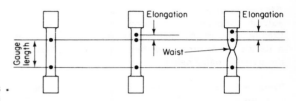

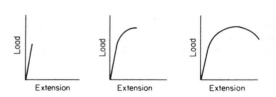

Fig. 4.5   The shape of the load-extension curve and the final shape of the test piece may be used as a measure of ductility

In actual practice the determination of points A and B is difficult and since points A, B and C are close together the yield point C is usually used to represent all three.

The quantities that are obtained from a tensile test are:
1. The stress at the yield point, which is defined as load at yield point divided by the *original* cross-sectional area of the test specimen.
2. The ultimate tensile stress, which is the ultimate load divided by the *original* cross-sectional area of the test specimen.
3. The percentage elongation at fracture, which is

$$\frac{\text{final extension}}{\text{original gauge length}} \times 100$$

4. The percentage reduction in area, which is

$$\frac{\text{original cross-sectional area} - \text{final cross-sectional area}}{\text{original cross-sectional area}} \times 100$$

The ultimate tensile stress is used to compare the strengths of various materials but it is not of much practical use since equipment and machinery are never operated at breaking point. The value of the yield stress is more important and designers often work to 50% of the yield stress.

VALUES OF THE ULTIMATE TENSILE STRENGTH

Material	U.T.S. (MN/m²)
Cast iron	75 to 200
Mild steel	400
Brass (cast)	120
Phosphor bronze	400

From the table we can see that that cast iron and cast brass are comparatively weak in tension whilst mild steel and phosphor bronze are much stronger.

Example. In a tensile test on a standard test specimen (diameter 11·3 mm and gauge length 56·5 mm) the following results were obtained:
Yield load = 25 kN Ultimate load = 54·5 kN
Diameter at fracture = 7·03 mm
Length over gauge marks at fracture = 71·1 mm
Find (a) the yield stress (b) the ultimate tensile stress (c) the percentage elongation (d) the percentage reduction in area.
Original cross-sectional area = $\pi r^2 = \pi \times 5.65^2 = 100$ mm²

(a) Yield stress

$$= \frac{\text{yield load}}{\text{original c.s.a.}} = \frac{25\,000}{100}$$
$$= 250 \text{ N/mm}^2 \text{ or } 250 \text{ MN/m}^2$$

(b) Ultimate tensile stress

$$= \frac{\text{ultimate load}}{\text{original c.s.a.}} = \frac{54\,500}{100}$$
$$= 545 \text{ N/mm}^2 \text{ or } 545 \text{ MN/m}^2$$

(c) Percentage elongation

$$= \frac{\text{total extension}}{\text{gauge length}} \times 100$$
$$= \frac{71.1 - 56.5}{56.5} \times 100 = 25.9\%$$

(d) Percentage reduction in area

$$= \frac{\text{original c.s.a.} - \text{final c.s.a.}}{\text{original c.s.a.}} \times 100$$

final c.s.a. $= \pi \times 3.515^2 = 38.82$ mm²

Percentage reduction in area

$$= \frac{100 - 38.82}{100} \times 100 = 61.18\%$$

Proof stress. Some materials do not exhibit a yield point (see Fig. 4.4) and for these materials a *proof stress* is used instead of the yield stress. This is determined from the load-extension graph as follows: If the 0·2% proof stress is required then AB (Fig. 4.6) is marked off to represent 0·2% of the gauge length of the test specimen. For instance, if the gauge length is 56·5 mm then

$$AB = \frac{0.2}{100} \times 56.5 = 0.113 \text{ mm. The line BC is then drawn}$$

parallel to straight portion of the graph. The load at the point C is found and the corresponding proof stress is calculated by dividing the load at C by the original cross-sectional area.

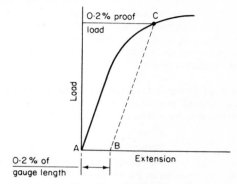

Fig. 4.6 Proof load

Example. The results below were obtained during a tensile test using a test piece with a diameter of 15·96 mm and a gauge length of 80 mm. Determine the 0·2% proof stress.

Load (kN)	10	20	30	35	40
Extension (mm)	0·02	0·04	0·06	0·07	0·082
Load (kN)	45	50	55	60	
Extension (mm)	0·10	0·175	0·31	0·45	

$$0.2\% \text{ of gauge length} = \frac{0.2}{100} \times 80 = 0.16$$

The graph is drawn in Fig. 4.7 and from the graph the 0·2% proof load is found to be 53·6 kN. Hence 0·2% proof

$$\text{stress} = \frac{0.2\% \text{ proof load}}{\text{original c.s.a.}}$$
$$= \frac{53\,600}{200} = 268 \text{ N/mm}^2 \text{ or } 268 \text{ MN/m}^2$$

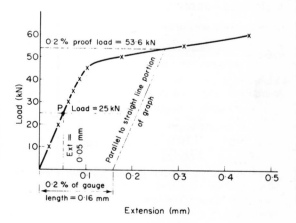

Fig. 4.7

Ductility. The percentage elongation and the percentage reduction in area are measures of the ductility of the material under test. For ductile materials the percentage elongation varies between 15% and 75% whilst for brittle materials it may be as low as 2%.

The shape of the load-extension curve and the final shape of the test piece (Fig. 4.5) will often show whether a metal is ductile or not.

Modulus of elasticity. It has been pointed out that up to the limit of proportionality the load (or stress) is proportional to the extension (or strain). That is,

$$\frac{stress}{strain} = constant$$

This constant is called *Young's modulus of elasticity* and it is usually denoted by the symbol E. Thus

$$E = \frac{stress}{strain}$$

since strain has no unit, the unit of E is the same as the unit of stress, usually meganewtons per square metre (MN/m^2). Some values of E are given below.

VALUES OF E

Material	Value of E (MN/m^2)
Steel	200 000
Cast iron	90 000 to 120 000
Brass	70 000 to 90 000
Bronze	100 000
Aluminium	60 000 to 85 000
Copper	104 000

Example. A mild steel bar 15·96 mm diameter and 80 mm long extended 0·060 mm under a tensile load of 30 kN. Calculate the value of E.

Applied load = 30 000 N

cross-sectional area of bar = $\dfrac{\pi \times 15 \cdot 96^2}{4}$ = 200 mm²

stress = $\dfrac{\text{applied load}}{\text{cross-sectional area}}$ = $\dfrac{30\,000}{200}$ = 150 N/mm² or 150 MN/m²

Original length = 80 mm
Alteration in length = 0·060

strain = $\dfrac{\text{alteration in length}}{\text{original length}}$ = $\dfrac{0 \cdot 060}{80}$ = 0·000 75

$E = \dfrac{stress}{strain} = \dfrac{150}{0 \cdot 000\,75}$ = 200 000 MN/m²

To find Young's modulus of elasticity from a test only the straight-line portion of the load-extension or stress-strain graph is used. The method is shown in the following example.

Example. During a tensile test on a bronze specimen with a diameter of 15·96 mm and a gauge length of 80 mm the following results were obtained.

Load (kN)	10	20	30	35	40	45	50	55
Extension (mm)	0·04	0·08	0·12	0·14	0·164	0·20	0·35	0·62

Find the value of Young's modulus of elasticity for bronze.

A point P is chosen on the straight-line portion of the load-extension curve (Fig. 4.8) and the load and

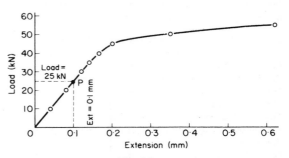

Fig. 4.8

corresponding extension at P determined. At P the load is 25 kN and the extension is 0·1 mm. Since the cross-sectional area for a test piece 15·96 mm its cross-sectional area is 200 mm².

stress at P = $\dfrac{25\,000}{200}$ = 125 N/mm² or 125 MN/m²

strain at P = $\dfrac{0 \cdot 1}{80}$ = 0·00125

$E = \dfrac{stress}{strain} = \dfrac{125}{0 \cdot 00125}$ = 100 000 MN/m²

Tensile test for plastics. This is conducted in the same way as for metals but the test piece (Fig. 4.9) is made

Fig. 4.9 Test piece for plastic materials

from sheet material. The type of load-extension curve obtained depends upon the temperature of the specimen and the straining rate (Fig. 4.10). Both the temperature

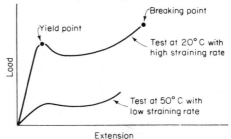

Fig. 4.10 Load-extension curves for polyethylene

(Fig. 4.11) and the straining rate have an effect on the tensile strength of the material (see table below).

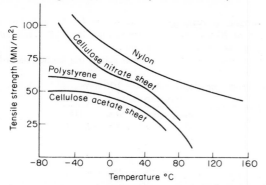

Fig. 4.11 Effect of temperature on the tensile strength of thermoplastic materials

EFFECT OF STRAINING RATE ON TENSILE STRENGTH OF POLYETHYLENE

Straining rate (mm/min)	Tensile strength (MN/m²)
150	18
300	19
450	20
750	22

For some plastics (polyethylene for example) the term 'modulus of elasticity' can be misleading since the stress obtained during a test depends upon the straining rate. Also, although the material is elastic in that the specimen will return to its original size when the load is removed this does not occur immediately (as happens with metals) but over a period of time which is related to the time the material was stressed. For such plastics no single value of E can be stated and the value obtained from a test should be regarded only as a guide since it will apply only to conditions similar to those of the test.

Compressive forces. Loads which tend to shorten rather than extend a material are called compressive loads. Compressive stresses and strains are calculated in exactly the same way as tensile stresses and strains, that is

$$\text{compressive stress} = \frac{\text{compressive load}}{\text{cross-sectional area}}$$

$$\text{compressive strain} = \frac{\text{amount of compression}}{\text{original length}}$$

The value of the modulus of elasticity is the same in compression as in tension.

Example. A brass bush, 25 mm O/D and 20 mm I/D is used as a spacer and it is not to compress by more than 0·01 mm when loaded. If it is 50 mm long find the maximum load it can carry. (E for brass is 80 000 MN/m²).

$$\text{Maximum strain} = \frac{0·01}{50} = 0·0002$$

$$E = \frac{\text{stress}}{\text{strain}}$$

$$\text{or stress} = E \times \text{strain}$$
$$= 80\,000 \times 0·0002$$
$$= 16 \text{ MN/m}^2 \text{ or } 16 \text{ N/mm}^2$$

$$\text{stress} = \frac{\text{load}}{\text{cross-sectional area}}$$

$$\text{or load} = \text{stress} \times \text{cross-sectional area}$$
$$= 16 \times \pi \,(12·5^2 - 10^2)$$
$$= 16 \times 176·8 = 2829 \text{ N}$$

The maximum load the bush can carry is 2829 N.

Rigidity. A column or a strut may be strong enough to stand the compressive stress on it but it may fail because of lack of rigidity (Fig. 4.12). The rigidity of strut depends

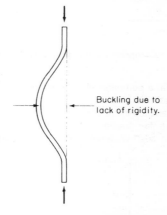

Buckling due to lack of rigidity.

Fig. 4.12 Lack of rigidity may cause a strut to fail

upon the ratio of diameter to length. The greater the length of the strut the larger its diameter needs to be.

Compression test. Machines used for compression tests are basically the same as those used for tensile tests and frequently the same machine is used for both tests. For metals the test piece is in the form of a short cylinder.

Ductile materials give a load-compression diagram similar to Fig. 4.13. The graph up to the yield point (point Y in the diagram) is very similar to that obtained during a tensile test. Beyond the yield point the diagram has an increasing upward slope due chiefly to the fact that the cross-sectional area of the specimen increases as its length decreases. The actual stress in the material does increase but much more slowly as shown by the dotted line in Fig. 4.13.

$$\left(\text{Note   actual stress} = \frac{\text{load}}{\text{actual c.s.a.}} \right)$$

Eventually the specimen becomes barrel shaped and surface cracks usually develop. No maximum or ultimate load can be found from the load-compression diagram and hence it is pointless to talk of an ultimate compressive

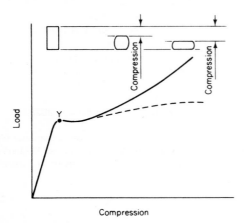

Fig. 4.13   Compression test on a ductile material

stress although a compressive yield stress may be found for ductile materials.

Brittle materials behave elastically up to a certain load but they then fail suddenly by splitting as shown in Fig. 4.14. For such materials the ultimate compressive

Fig. 4.14   How a brittle metal fails in compression

stress (the compressive strength or crushing strength) is defined as $\dfrac{\text{max. compressive load}}{\text{original c.s.a.}}$

The compression test is seldom used as an acceptance test for either metallic or plastic materials (except for cast iron) because it gives less information than a tensile test. Its use is generally restricted to building materials and for research into the properties of new materials.

Shearing forces. These are forces which tend to cause one layer of material to slide over another layer as shown in Fig. 4.15. The shear stress in a material is calculated as follows:

$$\text{shear stress} = \frac{\text{shearing force (or load)}}{\text{area resisting shear}}$$

Example. A plate 25 mm wide and 4 mm thick has a load of 20 kN applied as shown in Fig. 4.16. Calculate the shear stress in the material.
The material will tend to shear across the section aa and hence
area resisting shear = 25×4 = 100 mm
$$\text{Shear stress} = \frac{\text{shearing force}}{\text{area resisting shear}} = \frac{20\,000}{100}$$
$$= 200 \text{ N/mm}^2 \text{ or } 200 \text{ MN/m}^2$$

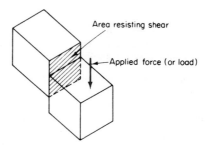

Fig. 4.15   Shearing force

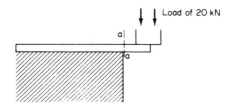

Fig. 4.16

Single and double shear. A bolt or a rivet may be placed in either single or double shear as shown in Fig. 4.17. For the rivet to fail in single shear it must fracture across the section aa and hence

$$\text{shear stress in single shear} = \frac{\text{shearing force}}{\text{c.s.a. of rivet}}$$

Fig. 4.17

For the rivet to fail in double shear it must fracture across both of the sections xx and yy and hence

$$\text{shear stress in double shear} = \frac{\text{shearing force}}{2 \times \text{c.s.a. of rivet}}$$

where c.s.a. stands for cross-sectional area.

Examples

1. A bolt 10 mm diameter is to hold two plates together (Fig. 4.17). If the load applied to the plates is 10 kN find the shear stress in the bolt.

    Area resisting shear = c.s.a. of bolt
    $$= \pi \times 5^2 = 78 \cdot 5 \text{ mm}^2$$

    Shear stress in bolt $= \dfrac{\text{shearing force}}{\text{area resisting shear}}$
    $$= \frac{10\,000}{78 \cdot 5}$$
    $$= 140 \cdot 1 \text{ N/mm}^2 \text{ or } 140 \cdot 1 \text{ MN/m}^2$$

2. Fig. 4.18 shows part of a link used in a machine tool. Calculate the maximum tensile stress in the members A and B and find the shear stress in the rivets.

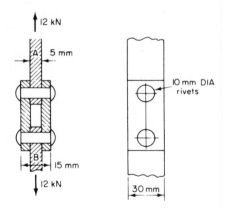

Fig. 4.18

The maximum tensile stress will occur across the holes and at these points

cross-sectional area = $5 \times (30 - 10) = 5 \times 20 = 100 \text{ mm}^2$
applied load = 12 kN = 12 000 N

$\therefore$ tensile stress in members A and B $= \dfrac{12\,000}{100}$
$$= 120 \text{ N/mm}^2$$
or
$$= 120 \text{ MN/m}^2$$

cross-sectional area of rivets $= \pi \times 5^2 = 78 \cdot 5 \text{ mm}^2$
The rivets are in double shear and hence

shear stress in the rivets $= \dfrac{12\,000}{2 \times 78 \cdot 5} = 76 \cdot 4 \text{ N/mm}^2$ or $76 \cdot 4 \text{ MN/m}^2$

Direct shear test. The methods used for direct shear tests on round and rectangular bars are shown in Fig. 4.19. Unfortunately the material is not placed in

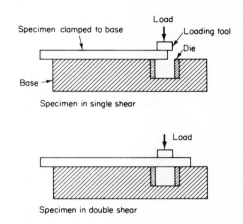

Fig. 4.19   Direct shear tests

pure shear because some bending must occur. The test is conducted at slow speed until the material fails and it is often performed in a tensile testing machine. Plates and rivets may be tested as shown in Fig. 4.20.

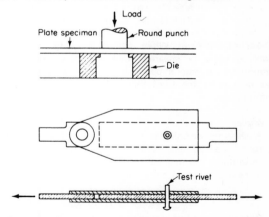

Fig. 4.20   Methods of performing shear tests on plates and rivets

For plastic sheet the test piece is mounted in a punch type shear fixture (Fig. 4.21) and a punch, generally 25 mm diameter, is pushed down at a speed of 1 mm/min until the material shears. The shear strength of the plastic is then calculated from

$$\text{shear strength} = \frac{\text{applied force}}{\text{area sheared}}$$

Example. A plastic sheet is tested by placing a test

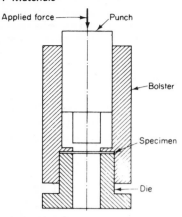

Fig. 4.21   Shear strength test for plastic sheet

As shown in Fig. 4.22 the area to be sheared is that round the circumference of the hole and hence

shearing area = circumference of hole × thickness
= $\pi \times 40 \times 7 = 880$ mm²

Force required = shearing area × shear strength
= $880 \times 300 = 264\,000$ N or 264 kN

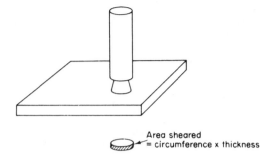

Fig. 4.22

specimen 5 mm thick in a shear fixture similar to Fig. 4.21. The punch was 25 mm diameter and the force applied to shear the plastic was 7·5 kN. Calculate the shear strength of the material.

Area sheared = circumference of hole × thickness
= $\pi \times 25 \times 5 = 235 \cdot 5$ mm²

Shear strength = $\dfrac{\text{applied load}}{\text{area sheared}} = \dfrac{7500}{235 \cdot 5}$
= 31·8 N/mm² or 31·8 MN/m²

Shearing material. When material is being cut on a blanking press or sheared on a guillotine the shear strength of the material becomes very important because this determines the force that must be applied to the cutting edge.

Examples

1. A shearing machine is used to crop off lengths of round bar 20 mm diameter. If the ultimate shear strength of the material is 160 MN/m² calculate the force needed to crop the bar.

   Area sheared = cross-sectional area of bar
   = $\pi \times 10^2 = 314 \cdot 2$ mm²

   Before the material will shear the stress in the material must be at least the shear strength, i.e. 160 MN/m² or 160 N/mm². Since

   shear stress = $\dfrac{\text{applied load}}{\text{area sheared}}$

   applied load = shear stress × area sheared
   = $160 \times 314 \cdot 2$
   = 50 300 N or 50·3 kN

   Thus a minimum force of 50·3 kN is required to crop the bar.

2. The shear strength of a plate is 300 MN/m². Calculate the force required to punch a hole 40 mm diameter in a plate 7 mm thick.

Twisting. When a shaft has a torque applied to it it tends to twist. The amount of twist depends upon the length of the shaft, its diameter, the amount of applied torque and the material from which the shaft is made.

Torsional stress. Shafts which transmit torque are subjected to torsional stress and the material encounters a shearing action. The shear stress in the material varies from a maximum at the surface of the shaft to zero at its centre. It is the maximum stress that is of interest and this is found by using the formula

$$\text{maximum shear stress} = \frac{16\,T}{\pi d^3}$$

where T = applied torque (N mm)
d = diameter of shaft (mm)

The maximum shear stress is then in newtons per square millimetre (N/mm²).

Examples

1. A boring bar 20 mm diameter has a torque of 60 Nm applied to it when it is cutting. Find the maximum shear stress in the material.

   $T = 60$ Nm = 60 000 Nmm

   Max. shear stress = $\dfrac{16 \times 60\,000}{\pi \times 20^3}$
   = 102 N/mm² or 102 MN/m²

2. The shear stress in a shaft is not to exceed 30 MN/m². The shaft is 100 mm diameter. What is the greatest torque that can be placed on it?

   Shear stress = 30 MN/m² = 30 N/mm²

   $30 = \dfrac{16\,T}{\pi \times 100^3}$

   $T = \dfrac{30 \times \pi \times 100^3}{16}$
   = 5 900 000 Nmm or 5900 Nm

The greatest torque that can be applied is 5900 Nm.

3. Two shafts are connected by a flanged coupling. The coupling is secured by 6 bolts 20 mm diameter on a pitch circle diameter of 150 mm. If a torque of 120 Nm is applied find the shear stress in the bolts.

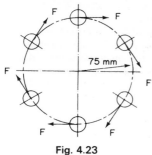

Fig. 4.23

As a result of the applied torque each bolt will have a shearing force F applied to it as shown in Fig. 4.23.

The total torque resisted by the bolts is $6 \times F \times 75$

$$\therefore 6 \times F \times 75 = 120 \times 1000$$

$$F = \frac{120 \times 1000}{6 \times 75} = 267 \text{ N}$$

In dealing with couplings it is assumed that the shear stress in the bolts is uniform. Hence

$$\text{shear stress} = \frac{267}{\pi \times 10^2} = 0.85 \text{ N/mm}^2 \text{ or } 0.85 \text{ MN/m}^2$$

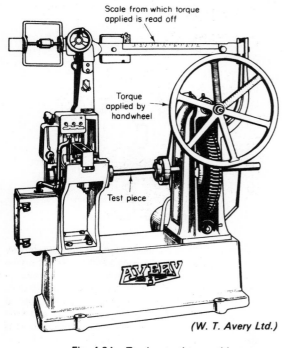

(W. T. Avery Ltd.)

Fig. 4.24   Torsion testing machine

Torsion test. The machine used for a torsion test may be similar to that shown in Fig. 4.24 where the torque is applied by the hand wheel, but larger machines are motor driven. In order to plot a torque-twist diagram it is necessary to measure the angle of twist and the optical twist meter (Fig. 4.25) is suitable for this purpose.

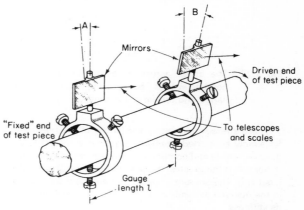

Fig. 4.25   Torsion twist-meter. Angle of twist is B−A. (from *Mechanical testing of materials* by A. J. Fenner, Newnes 1965)

Fig. 4.26 shows a typical torque-twist diagram. The graph is a straight-line up to the point Y at which point the metal at the surface of the specimen reaches its yield point. The material in the interior of the specimen will be still within the elastic limit because the maximum shear stress occurs at the surface of the bar. However, as the test proceeds the material in the interior will eventually reach the plastic stage.

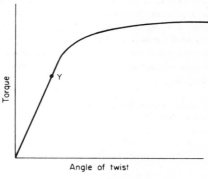

Fig. 4.26   Torsion test

Bending. When a beam bends, a tensile stress is set up in one flange and a compressive stress in the other. Hence the stress across the section is not uniform across the section of the material (see Fig. 4.27). The web is subjected to a shear stress.

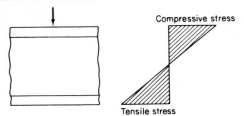

Fig. 4.27   The stress across a section of a beam in bending varies, being compressive at one flange and tensile at the other

These stresses vary along the length of the beam and for simply supported beams it may be assumed that the greatest tensile and compressive stresses occur where the deflection is the greatest. For cantilevers they occur at the fixed end. The amount of deflection and hence the stress in the material depends upon the length of the beam and its stiffness.

The stiffness of a beam depends upon the cube of its depth. Thus the deeper the beam the greater its stiffness:

a beam of depth d has a stiffness proportional to d^3
a beam of depth $2d$ has a stiffness proportional to $8d^3$.

Therefore by increasing the depth by 2 the stiffness is increased 8 times.

The deflection of beams is important in machine tools. The arbor of a milling machine has loads imposed upon it by the cutting forces at the cutter. In order to keep deflections to a minimum the length of unsupported arbor should be kept as short as possible and the maximum diameter of arbor used.

Flexure tests. These are used chiefly for testing cast iron and plastics. The test consists of placing the test specimen on two supports and applying a load at the centre until the specimen breaks (Fig. 4.28). The material is satisfactory if it withstands a minimum fracturing load. In the case of cast iron a minimum deflection may also be specified.

Bend tests. These are performed on ductile metals to give some indication of the capacity of the material to

deform in one particular direction without cracking. For round and rectangular bars the test is conducted as shown in Fig. 4.29(a), the bar being required to bend without cracking until the angle of bend, α, reaches a specified value.

The close bend test (Fig. 4.29(b)) is conducted by bending the test piece through 180° by using a former. The U so formed is then closed until the inner surfaces of the test piece are in contact. The material is satisfactory if the test can be completed without the metal cracking. For thin material, pressing the test piece into a block of soft lead by means of a former (Fig. 4.29(c)) is recommended.

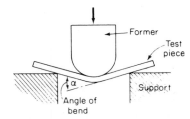

(a) Bend test for round and rectangular bar

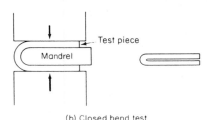

(b) Closed bend test

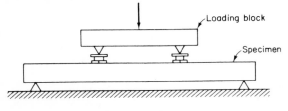

(c) Bend test for thin material

Fig. 4.29   Bend tests

Reverse bend tests are used on sheets and strip of very ductile material where the closed bend test is not severe enough. One end of the test piece is fixed in a vice or machine between mandrels having a specified radius. The projecting length of the test piece is then bent at right-angles to the fixed end first to one side and then to the other until the test piece breaks. The material is satisfactory if it withstands a specified number of bends.

Another test for the ductility of thin sheet is the

Fig. 4.28   The flexure test

Erichson cupping test. A spherical indenter is forced into the test specimen as shown in Fig. 4.30. The more ductile the metal the greater is the depth of the cup before a crack appears.

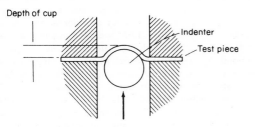

Fig. 4.30   The Erichson cupping test

Hardness tests. Many methods and machines for testing the hardness of materials have been developed but those in general use are the Brinell, Vickers Diamond, Rockwell and Shore scleroscope tests.

Brinell Test. A hard steel ball, usually 10 mm diameter, is pressed into the test specimen by applying a standard load which is usually 3000 kg. The ball is then removed and the diameter of the identation (Fig. 4.31) measured

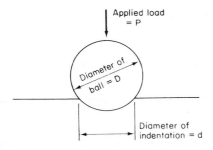

Fig. 4.31   The Brinell hardness test

by using a microscope. The hardness number may then be calculated from the relationship

$$B.H.N. = \frac{Applied\ load\ (kilograms)}{spherical\ area\ of\ the\ impression\ (square\ millimetres)}$$

$$= \frac{P}{1 \cdot 571\ D(D - \sqrt{D^2 - d^2})}$$

where D = diameter of ball (millimetres)
  d = diameter of the impression (millimetres)
and P = applied load (kilograms)

However, in practice, conversion tables such as the one below are used to find the Brinell Hardness Number.

BRINELL HARDNESS NUMBERS
Diameter of ball = 10 mm.   Load on ball = 3000 kg.

DIA of impression	BHN	DIA of impression	BHN
2·3	713	3·8	255
2·4	652	3·9	241
2·5	600	4·0	228
2·6	555	4·1	217
2·7	512	4·2	207
2·8	477	4·3	196
2·9	444	4·4	187
3·0	418	4·5	179
3·1	389	4·6	170
3·2	364	4·7	163
3·3	340	4·8	156
3·4	321	4·9	149
3·5	302	5·0	143
3·6	286	5·1	137
3·7	269	5·2	131

The standard 3000 kg load and 10 mm diameter ball are departed from only for tests on thin sheets and soft material and B.S. 240 provides for a number of different ball sizes and the corresponding loads. For very hard materials with a B.H.N. greater than about 550 a tungsten-carbide ball is used because the standard steel ball then tends to flatten.

The Brinell hardness test may be used to estimate other properties as follows:

1. *Tensile strength.* There is a relationship between hardness and tensile strength which, for carbon and alloy steels, is

    ultimate tensile strength (MN/m²) = B.H.N. × 3·4

2. *Machineability.* Materials having a B.H.N. greater than 300 are difficult to machine using ordinary tools. A material with a B.H.N. below 120 is soft and will tear under the cutting edge particulary during screw cutting, milling and broaching.

3. *Work hardening capacity.* The form of the impression gives an indication of the work hardening capacity of the material (Fig. 4.32).

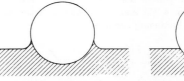

Piling up. The material will not work harden to any great extent when cold worked

Sinking. The material will work harden quickly when cold worked

Fig. 4.32   Form of the impression used as a measure of work hardening capacity

Vickers Diamond Test. The test is similar to the Brinell test except that the steel ball is replaced by a diamond indenter having the form of a square based pyramid with an angle of 136° between opposite faces. The indenter makes a square pyramidal impression and the diagonals of this impression are measured by means of a microscope which is usually part of the testing machine (Fig. 4.33).

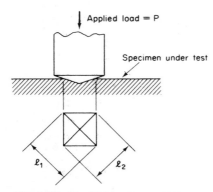

Fig. 4.33   The Vickers Diamond Test

The Vickers Pyramid Number is found from the relationship

$$V.P.N. = \frac{Applied\ load\ (kilograms)}{surface\ area\ of\ impression\ (square\ millimetres)}$$

$$= \frac{2P}{l^2\ cosec\ \dfrac{\theta}{2}}$$

where P = applied load (kilograms)
l = mean length of diagonals (millimetres)
and θ = angle of the diamond pyramid.
Since θ = 136°, the relationship becomes

$$V.P.N. = \frac{1 \cdot 854P}{l^2}$$

where $l = \frac{1}{2}(l_1 + l_2)$ as shown in Fig. 4.33.

As with the Brinell test the shape of the indentation gives a measure of the work hardening capacity of the material under test. Although the shape of the indentation is nominally square sometimes other shapes are obtained (see Fig. 4.34) and these may be used as a measure of work hardening capacity.

Material will not work harden to any great extent   Normal indentation   Material will work harden readily

Fig. 4.34   Shapes of indentations obtained during a Vickers Diamond Test.

The Rockwell Test. This test also uses indenters which are pressed into the surface of the test piece. It differs from the Brinell and Vickers tests in that the measurement of hardness is based on the depth of penetration of the indenter. The test is very quick because the hardness number is read directly on the dial of the machine. Two types of indenter are used— a conical diamond of 120° included angle for hard materials and a steel ball for softer materials. When the diamond is used the hardness number is referred to as Rockwell C and when the ball is used as Rockwell B.

In order to eliminate spring and backlash in the machine a minor load of 10 kg is first applied and the dial set to read zero. The major load is now applied and then removed leaving the minor load still acting. The hardness of the material is then determined by the increase in the depth of penetration caused by the major load, that is d in Fig. 4.35. The harder the material the

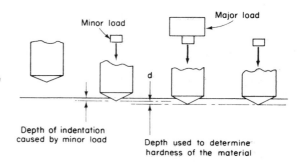

Fig. 4.35   The Rockwell Hardness Test

smaller the depth of penetration. The hardness number is given by the expression

Rockwell Hardness Number = K − 500d

K is given the value of 100 when a diamond indenter is used and 130 when using a steel ball.

The Rockwell test is often used for checking the hardness of plastics but the material must be at least 6 mm thick. The indenter used is a steel ball of larger diameter than the one used for metal. Diameters up to 12·7 mm are used giving scales denoted by the letters K, E, M, L and R (in descending order of hardness). Values obtained with one set of balls cannot be directly compared with values from another. Rockwell hardness can differentiate relative hardnesses of different types of a given plastic but it should not be used to compare hardnesses of various kinds of plastic. This is because elastic recovery is involved as well as hardness. It should be noted that Rockwell hardness is not an index of wear or abrasive resistance. Polystyrene, for instance, has a high Rockwell hardness but a low resistance to abrasion.

Shore Scleroscope. In each of the hardness tests previously described the measurement of hardness is based on the dimensions of the indentation. The Shore Scleroscope test measures hardness by allowing a diamond tipped hammer to fall on the test piece and measuring the height of the rebound (Fig. 4.36). The

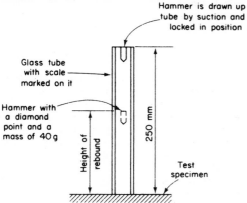

Fig. 4.36   Shore scleroscope

test works on the principle that for hard metals the amount of rebound is greater than for soft metals.

The advantages of the scleroscope are

1. It is small enough to be carried to the specimen unlike the Brinell, Vickers and Rockwell machines. It may therefore be used for checking the hardness of large components in situ.
2. The mark made by the hammer of scleroscope is very small and causes negligible damage to the test piece.

The scale of the scleroscope is graduated into 140 divisions a rebound of 100 divisions being equivalent to the hardness of hardened high carbon steel.

Comparison of hardness tests. No precise relationship exists between the various types of hardness numbers. However, approximate relationships have been found by performing tests on the same materials using the various methods. The table below will give some idea of comparative hardness numbers.

Material	Brinell	Vickers	Rockwell	Shore sclero-scope
Aluminium alloy	100	—	B 57	—
Mild steel	130	—	B 73	20
Annealed high carbon steel	230	241	C 20	34
Cutting tools	650	697	C 60	80

Impact tests. These check the toughness of a material by striking the test specimen with a controlled blow and measuring the amount of energy needed to bend or break the specimen. This amount of energy is a measure of the toughness of the metal. There are two kinds of test in common use, namely the Charpy and Izod tests.

Izod test. Details of this test are given in Fig. 4.37 where the test specimen is supported vertically as a cantilever. Usually the test machine has a capacity of about 160 J but other sizes are available.

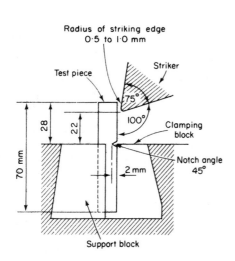

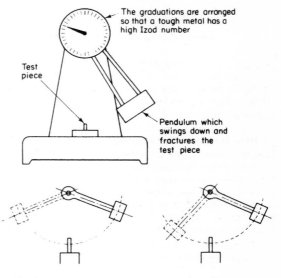

Fig. 4.37   The Izod impact test

Charpy test. Details are given in Fig. 4.38. The main difference to the Izod test is that the specimen is supported horizontally at both ends i.e. as a beam. The

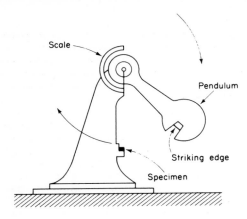

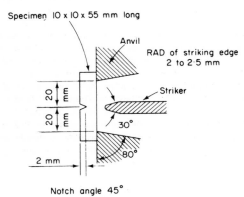

Fig. 4.38   The Charpy impact test

Charpy test is often preferred to the Izod because in the Izod test the specimen is stressed in the region of the notch by the vice. The Charpy machine may be obtained in a variety of sizes but a usual size has a capacity of about 300 J.

Brittle fracture. Surface irregularities are a potential source of cracks and parts made from ductile materials sometimes fail because a crack, originating with a small surface defect, spreads rapidly throughout the material. This type of failure is called brittle fracture because the material, although ductile, has failed in a way associated with brittle materials. The notch in the impact test specimen reproduces the conditions under which brittle fracture is likely to occur and hence an impact test indicates the resistance of the material to brittle fracture.

Impact test for plastics. Either the Izod or the

Charpy test may be used and these tests are useful for comparing various types or grades of plastic. Some materials, notably nylon and acetal-type plastics, are notch sensitive in that they register low values on the notched Izod impact test. One way of overcoming notch sensitivity is to use a tensile impact test in which the test piece is held clamped at one end whilst a projecting bar is clamped at the other end. The projecting bar is then struck by a pendulum so that the specimen is subjected to tensile impact.

Exercise 4.1

1. A tensile force of 20 kN is applied to a member with a cross-sectional area of 100 mm². Calculate the stress in the member.
2. A tie bar 50 mm×30 mm in section has a load of 300 kN applied to it. Find the stress in the bar.
3. A bar 3 m long has a tensile force applied to it and as a result it extends by 0·48 mm. Calculate the strain.
4. A brass wire 600 mm long extends by 0·20 mm when a load is hung on its end. Calculate the strain in the wire.
5. The following results were obtained as a result of a tensile test on a specimen 80 mm gauge length and 15·96 mm diameter.

Load (kN)	10	20	30	40	50	60
Extension (mm)	0·024	0·046	0·074	0·096	0·120	0·144
Load (kN)	70	80	90	100	110	120
Extension (mm)	0·166	0·194	0·216	0·240	0·300	0·566

The specimen broke at 150 kN and its final diameter was 11·05 mm. Plot a graph of load against extension and from it find (a) the 0·2% proof stress; (b) the percentage reduction in area.

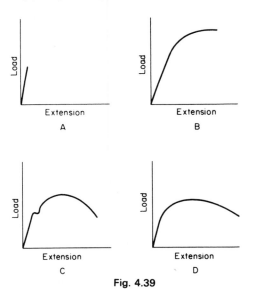

Fig. 4.39

6. In a tensile test the standard specimen used had a diameter of 22·56 mm and a gauge length of 113 mm, and the following results were obtained:
Yield load = 100 kN; Ultimate load = 220 kN; diameter at fracture = 11·58 mm; length over gauge marks at fracture = 142·5 mm. Calculate (a) the yield stress (b) the ultimate tensile stress (c) the percentage elongation (d) the percentage reduction in area.

7. A bar of steel whose cross-sectional area is 200 mm² has a tensile load of 40 kN applied to it. If E for steel is 200 000 MN/m² calculate the extension of the bar if its original length was 500 mm.

8. Fig. 4.39 shows the results of tests on four different metals, all the diagrams being drawn to the same scale. State which is (a) the most brittle (b) the most ductile and (c) the strongest metal.

9. A mild steel test specimen (11·28 mm diameter and gauge length 56 mm) extended 0·042 mm under a tensile load of 15 kN. Calculate the value of E.

10. The results below were obtained during a tensile test using a test piece with a diameter of 15·96 mm and a gauge length of 80 mm. Find the value of Young's modulus of elasticity

Load (kN)	5	10	15	17·5	20
Extension (mm)	0·01	0·02	0·03	0·035	0·041

Load (kN)	22·5	25	27·5	30
Extension (mm)	0·050	0·088	0·15	0·23

11. A steel bar 22 mm diameter and 60 mm long is not to compress by more than 0·01 mm when loaded. Find the maximum load it can carry if E for steel is 200 000 MN/m².

12. A bolt 20 mm diameter has a shearing force of 20 kN applied to it. Calculate the shear stress in the metal (a) if the bolt is placed in single shear (b) if it is placed in double shear.

13. Calculate the force required to punch a hole 12 mm diameter in a plate 5 mm thick if the ultimate shear stress of the metal is 200 MN/m².

14. A cropping machine is used to cut off lengths of bar 15 mm diameter. If the ultimate shear strength of the material is 220 MN/m² calculate the force required to crop the bar.

15. A shaft 40 mm diameter has a torque of 10 Nm applied to it. Find the maximum shear stress in the material.

16. A flanged coupling is used to connect two shafts. The coupling is secured by 5 bolts, 20 mm diameter, on a 200 mm pitch circle diameter. If a torque of 150 Nm is applied to the coupling find the stress in the bolts.

17. Why is a tensile test often preferred to a compression test? An ultimate compressive stress is never quoted for ductile metals. Why?

18. What is the purpose of bend tests? Name two such tests and state how the tests are conducted.

19. Show how the Brinell Hardness test may be used to estimate (a) tensile strength (b) machineability (c) work hardening capacity.

20. What advantages does the shore Scleroscope have over other hardness tests?

21. How are plastic materials tested for hardness?

22. Name two impact tests and by using clear diagrams show how the two tests differ. What is the purpose of the impact test?

4.2   Materials used in engineering

Aluminium. Pure aluminium has low strength but because of its excellent ductility and resistance to corrosion it is used in both the rolled and extruded form. It can be strengthened by cold working and even more by alloying.

Wrought aluminium alloys—not heat treated. These consist of the following:
Aluminium alloys containing 1 to 1·25% manganese.
Aluminium alloys containing 2 to 7% magnesium.

In the work hardened condition these alloys have adequate strength, rigidity and corrosion resistance. The required mechanical properties (i.e. strength, hardness, etc.) are produced by cold working after the last annealing operation.

Wrought aluminium alloys— heat treatable. These consist of the following:
Aluminium alloys containing between 0·7 and 1% magnesium and between 0·5 and 1% silicon.
Aluminium alloy containing 1·5% copper; 1% magnesium and 1% silicon.
Aluminium alloys containing 4·25% copper with magnesium, silicon and manganese (this alloy is known as duralumin).
Aluminium alloys containing 2·25% copper, 1 to 1·5% magnesium, 1% nickel with silicon and iron.

The heat-treatable alloys are stronger than the non-heat-treated types but their corrosion resistance is not as good and hence rolled sheet is often protected by coating the surface with a thin skin of aluminium. The clad-sheet is therefore high strength metal with high corrosion resistance. Duralumin coated with aluminium is called Alclad.

When dealing with aluminium the term heat treatment usually refers to the processes of solution treatment and precipitation treatment which are used to strengthen the heat treatable alloys.

Solution treatment consists of raising the temperature of an alloy (between 460 °C and 530 °C depending upon the alloy) to a point where the alloying elements are in solid solution in the aluminium and then quenching rapidly, usually in water. The alloy remains soft for a period during which time it may be cold worked but it then gradually hardens, the process being completed in about five days at room temperature. The natural hardening process is called age hardening.

Some alloys age so slowly that the process of hardening is speeded up by using the *precipitation treatment* which consists of heating the alloy to a temperature under 200 °C for a suitable period. Under these conditions artificial ageing is promoted.

Aluminium casting alloys. Foundry alloys are also divided into those which are heat treatable and those which are not but the non-heat treatable alloys cannot be work hardened as the wrought alloys can. Further sub-divisions occur because some alloys are more suitable for particular casting methods than others. The principle types are:
Aluminium—silicon (5 to 12% silicon)
Aluminium—magnesium (5 to 10% magnesium)
Aluminium—copper (4·5 to 10% copper)
Y—alloy (4% copper, 2% nickel and 1·5% magnesium).

For the heat treatable types solution treatment followed by precipitation treatment produces the highest strength.

British Standards for wrought aluminium alloys. These are as follows:
BS 1470 Plate, sheet and strip
BS 1471 Drawn tube
BS 1472 Forgings
BS 1473 Material for rivets, bolts and screws
BS 1474 Bars, extruded round tubes and sections
BS 1475 Wire

The form in which the material, referred to in the BS specifications above, is available is indicated by a letter as follows:
B Bolt and screw stock
C Clad sheet and strip
E Bar and sections
F Forgings
G Wire
P Plate
R Rivet stock
S Sheet and strip
T Drawn tube
V Extruded tube and hollow sections

The specifications are arranged in numerical order corresponding approximately to increasing strength and hardness. No. 1 is used for pure aluminium, Nos. 2 to 8 for non-heat treatable alloys and 9 to 30 for heat-treatable alloys.

Non-heat treatable alloys are prefixed with the letter N and are available in a range of tempers as follows:

O   Annealed
OD   Annealed and lightly drawn
M   As manufactured (rolled, forged etc.)
$\frac{1}{4}$H   Quarter-hard   ⎫ Degrees of temper pro-
$\frac{1}{2}$H   Half-hard   ⎬ duced by cold working on
$\frac{3}{4}$H   Three-quarters-hard ⎧ sheet, strip, plate, drawn
H   Full hard   ⎭ tube and rivet stock

An example might be NP7-$\frac{1}{4}$H which denotes non-heat treatable alloy No. 7 in the form of plate with quarter-hard temper.

Heat-treatable alloys are prefixed with the letter H and the condition of the supplied metal is indicated by one of the following:

O   Annealed
M   As manufactured (forged, rolled etc.) with no subsequent heat treatment
OD   Annealed and slightly drawn
W   Solution treated and will respond to precipitation treatment
T   Solution treated and requiring no precipitation treatment
WP   Solution and precipitation treated
P   Precipitation treatment only

An example might be HE30-WP which denotes heat-treatable alloy No. 30 in the form of bar or section which has been solution and precipitation treated.

Details of some of the wrought alloys available are given in Table 1 (see p. 100).

British Standards for aluminium casting alloys (BS 1490). All the alloys covered by this specification are prefixed with the letters LM. The numbers that follow these letters give the number of the alloy e.g. LM 6, LM 12 etc. When the material is referred to without a suffix it denotes that it is in ingot form. Other conditions are denoted by the following suffixes:

M   As cast
P   Precipitation treated
W   Solution treated
WP   Solution and precipitation treated

For example, LM 12 M would be quoted when a casting in alloy No. 12 is required in the as cast condition (that is with no heat treatment.)

Details of some of the casting alloys available are given in Table 2 (p. 101).

Alloys of copper. The three main copper alloys are
Brass (copper-zinc alloys)
Bronze (copper-tin alloys)
Aluminium bronze (copper-aluminium alloys)

The properties and compositions of some of the brasses and bronzes are given in Table 3 (p. 102).

Aluminium bronze may be obtained as plate, rods, sections and forged products and as tube, sheet, strip and wire. The mechanical properties of an aluminium bronze whose composition is copper 95% aluminium 5% are as follows:

	Annealed	Work Hardened
0·1% Proof stress	110 MN/m²	460 MN/m²
Ult. tensile stress	340 MN/m²	620 MN/m²
Percentage elongation	70	10
Vickers Hardness	85	180

This bronze can be cold worked to give high strength products with a good corrosion resistance.

Magnesium. Magnesium alloys are light, possess high strength and have excellent machineability. They can be obtained as castings (sand, gravity die, pressure die etc.) and as forgings, extrusions and sheet. Cast magnesium alloys make good bearing materials under light loads and may be used without bushes. Under heavy loads, however, cast steel or yellow metal bushes are needed. Wrought alloys have excellent machineability and are easy to weld. An idea of the properties of cast and wrought alloys may be obtained from the data below:

Property	Cast alloy		Wrought alloy Extruded bars
	as cast	fully H.T.	
0·1% Proof stress (MN/m²)	60	80	150
U.T.S. (MN/m²)	120	190	280
Elongation %	1	1	8
Hardness (B.H.N.)	50	70	65

Magnesium alloys are susceptible to corrosion being about the same as mild steel in this respect. Hence care must be taken to avoid very humid conditions particularly when the magnesium alloys are in contact with other metals when galvanic corrosion will occur.

Zinc alloys. Zinc alloy die castings are used for a large range of components ranging in size from a few grammes to 20 kg and are used as levers, gear wheels, instrument frames etc. The main alloying element is aluminium, about 4% being used, but small amounts of magnesium and iron are also incorporated in the mix. The alloys have an ultimate tensile strength of about 280 MN/m² with a percentage elongation of 10%.

Titanium. This is one of the newer engineering metals and it has only been available since 1948. The outstanding advantages of titanium alloys are their high strength, low density and high corrosion resistance. The properties of the alloys vary with composition but Titanium 314 (annealed rod) which has a composition of aluminium 4%, manganese 4% and titanium 92% has the following mechanical properties: 0·1% proof stress = 1000 MN/m², U.T.S. = 1050 MN/m² and percentage elongation = 19%.

Most of the titanium produced is used in the aerospace industry being used mainly for highly stressed components operating at elevated temperatures e.g. gas-turbine compressor discs, blades and castings. The machineability of titanium alloys is similar to that of alloy steels of the same strength.

Lead alloys. Lead-antimony alloys containing between 6 and 12% antimony are used in the manufacture of castings for the chemical industry when a considerable degree of hardness is required.

Lead-tin-antimony alloys are used for bearings. Some idea of the properties of lead and its alloys may be obtained from the table below.

PROPERTIES OF LEAD AND ITS ALLOYS

Property	Lead	Antimony–lead		Tin–lead	
		4% Sb	8% Sb	20% Sn	50% Sn
U.T.S. (MN/m²)	14	80*	86*	40	42
Hardness (BHN)	3–6	24*	26*	11	14
Density (kg/m³)	11 300	11 000	10 700	10 200	8 900

Sb = antimony   Sn = tin   *Annealed and aged for 1 day

Cast iron. Grey cast iron has graphite in flake form distributed throughout the metal. The graphite flakes vary in size and quantity and generally the fewer the flakes the higher the strength. BS 1452 specifies seven grades of cast iron numbered Grade 10, 12, 14, 17, 20, 23 and 24, grade 10 being the weakest and grade 24 the strongest. Some idea of the properties of grey cast iron can be obtained from the table below.

PROPERTIES OF GREY CAST IRON

Grade	Tensile strength (MN/m²)	Comp. strength (MN/m²)	Elonga-tion %	Hard-ness (BHN)
10	150	620	0·5	160
17	260	770	0·5	190
24	370	1230	0·5	240

Grey cast iron is used for crankshafts, cylinder blocks, gear wheels, pulleys, machine tool parts, handling equipment etc.

Nodular iron has its graphite aggregated into nodules or spheroids and this gives it properties which are much closer to those of steel. The material is often known as spheroidal graphite (SG) iron. Nodular irons are specified in BS 2789 which lists six grades. The properties of three of these grades are given in the table below.

PROPERTIES OF SPHEROIDAL–GRAPHITE CAST IRON

Type	U.T.S. (MN/m²)	Elongation %	Hardness (BHN)
SNG 370/17	370	17	140
SNG 500/7	490	7	180
SNG 700/2	730	2	250

Some of the uses of SG iron are levers, crankshafts, baseplates, frames for machinery, gears, wheels, bearing housings, cams and followers, handling plant and valve bodies.

Malleable irons consist of three main types known as whiteheart, blackheart and pearlitic which are specified in BS 309, BS 310 and BS 3333 respectively. Their mechanical properties are given in the table below.

PROPERTIES OF MALLEABLE IRONS

Type	U.T.S. (MN/m²)	Yield point (MN/m²)	Elonga-tion %	Hard-ness (BHN)
Whiteheart	340	200	6	240
Blackheart	300	180	10	140
Pearlitic	450	—	4	220

Malleable irons can be cast into complex shapes and their machineability is better than any other ferrous metal with a comparable hardness. Some grades have a high resistance to atmospheric corrosion, others have high impact strength and others are capable of being surface hardened after machining. Malleable irons are used for wheel hubs, cam-shaft brackets, couplings for rolling stock, cable supports, vice sideplates, gears and in any situation where ductile components in the form of castings are required.

Many types of *carbon and low alloy steels* produced in wrought form can be made into steel castings. Their mechanical properties are not noticeably affected by casting as can be seen in the table below.

PROPERTIES OF CARBON AND LOW ALLOY STEEL CASTINGS

BS spec.	Type	U.T.S. (MN/m²)	Yield point (MN/m²)	Elongation %	Hardness (BHN)
592A	carbon steel	430	230	22	—
1398A	carbon-molybdenum	460	280	20	—
1456A	1·5% manganese	540	320	18	—
1458	low alloy steel	700	490	15	220

These castings are used for general engineering purposes, ships and rolling stock. The castings are usually supplied in the annealed condition and may be finished by normalising or hardening and tempering.

The *cast high alloy steels* have a wide range of compositions, see Table 4 (p. 103). Generous machining allowances are usually allowed on sand castings owing to uncertainty in the amount of contraction and the lack of precision in the sand mould. None of the high alloy steel castings is easy to machine and with the austenitic steels drilling and milling is difficult. The manganese wear resisting steels require special techniques if they are to be machined at all.

Carbon steels. These are essentially alloys of iron and carbon together with varying amounts of other elements. The steels used in general engineering are specified in BS 970 and the properties and uses of these steels are given in Table 5 (p. 104).

Low and medium alloy steels. The mechanical properties of carbon steel may be improved by alloying with various elements. Table 6 (p. 104) gives the composition and properties of a selection of alloy steels.

Stainless steels contain more than 10% chromium with or without other alloying elements. They are available as wire, tube, plate, strip, bar, sections (extruded, rolled or drawn), sheet, forgings and castings. Some of the stainless steels available are given in Table 7 (p. 105).

High strength steels. These are structural steels with tensile strengths of 1500 MN/m² or more and reasonable ductility. The composition and properties of some of the H.T. steels available are given in Table 8 (p. 106).

Free cutting steels. Steels which have constituents to make them more easily machineable are called free cutting steels. The commonest elements used for this purpose are sulphur and lead. A typical sulphurised free cutting steel has a composition of 0·30% carbon, 0·25% silicon, 1·30% manganese and 0·18% sulphur. It has a tensile strength of 620 MN/m² and an elongation of 15%.

Thermoplastics. These materials soften when heated and harden upon cooling. The properties and uses of some of the thermoplastics are given in Table 9 (p. 106). They are obtainable as mouldings, extrusions and sheet.

Thermosetting plastics. These undergo a chemical change and set solid after being melted to a liquid state. They cannot be resoftened and further heating results only in a chemical breakdown, not melting. The properties and uses of some of the thermosetting plastics are given in Table 10 (p. 107). They may be obtained as mouldings etc.

Relative costs. Some idea of the relative costs of plastic and metallic materials can be obtained from the table below. It is based on the cost per cubic centimetre.

Material	Relative cost
PVC polymer	0·54
Polyethylene	0·54
Rigid PVC	0·62
Polypropylene	0·76
Polystyrene	0·59
Urea formaldehyde powder	1·00
Nylon powder	3·14
Steel (ingot)	9·8
Steel (thin sheet)	17·0
Copper	13·4
Aluminium	2·3
Tin	39·0

Nitriding. Steel can be given a hard, wear-resisting surface by using nitrogen as the hardening agent. This surface hardening process, called nitriding, consists of heating the steel at a temperature of 500 °C for up to 100 hours in a gas-tight container through which ammonia gas is circulated. The ammonia partly dissociates into nitrogen and hydrogen at the surface of the steel and the nitrogen diffuses into the surface to form hard nitrides.

Iron nitride is not the hardest nitride and it also tends to be excessively brittle, therefore plain carbon steels are not usually nitrided. Special alloy steels are used, for example, those of the Nitralloy type which contain 0·2–0·5% carbon, 1·5% chromium, 1% aluminium and 0·2% molybdenum. Hard nitrides of chromium and aluminium are formed at the surface and a surface hardness of the order of 1000 Vickers Pyramid Number may be obtained. The depth of case obtained depends upon the duration of the process and as shown in Fig. 4.40 a case 0·7 mm deep is obtained after 100 hours.

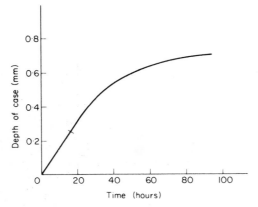

Fig. 4.40   Relation between depth of case and duration of the nitriding process

Parts to be nitrided are given heat treatment prior to nitriding in order to obtain a tough core. The full sequence for a Nitralloy steel would be as follows:

(1) Harden by oil quenching from 900 °C.
(2) Temper at 650 °C.
(3) Rough machine.
(4) Relieve machining stresses by heating to about 550 °C for 5 hours.
(5) Finish machine.
(6) Nitride at 500 °C.

Spindles, brake drums, moulds for plastics, crankshafts etc. are typical examples of parts which are nitrided. Note that the parts can be finished-machined before nitriding because no quenching is required and hence they will not suffer from distortion as a result of quenching.

Nitriding is an ideal way of producing a hard surface but it is economical only for large-scale production because of the high initial cost of the nitriding plant. It has a number of advantages including: (a) a very hard surface is obtained, (b) the case hardness is retained up to 500 °C (carburised cases begin to soften at 200 °C), (c) the nitrided case resists corrosion, and (d) distortion and cracks due to quenching are avoided because heat treatment is carried out prior to nitriding.

Flame hardening. Steels containing 0·4% to 0·6% carbon are heated to the hardening temperature by using a movable oxy-acetelene flame. They are then quenched

by using a water spray attached to the oxy-acetelene torch (Fig. 4.41). With a 0·45% carbon steel a surface hardness of about 600 VPN is obtained and the depth of case is usually about 3 mm.

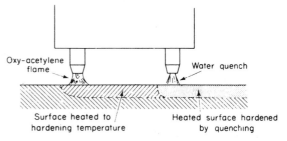

Fig. 4.41   Flame hardening

The process is used for hardening gears, cams, sprockets, spindles etc. and before hardening they should be stress relieved. After flame hardening the steel is usually tempered and this can be done by flame heating if required.

Induction hardening. The component to be hardened is surrounded by a perforated induction coil as shown in Fig. 4.42. The induced electric current heats the surface

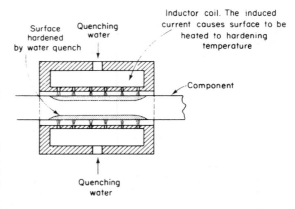

Fig. 4.42   Induction hardening

of the component to the hardening temperature in about 5 seconds. When this temperature is reached the current is switched off and quenching is then carried out by spraying with water through the holes in the induction coil.

This is a production method of surface hardening and steels with about 0·45% carbon are most suitable for induction hardening. A case about 3 mm deep is obtained and by shaping the induction coil the heating can be localised.

Tempering. Steels are usually tempered after hardening to improve their toughness and ductility. Tempering consists of heating the metal to the tempering temperature and quenching, although for the plain carbon steels the rate of cooling is unimportant.

The tempering temperature depends upon the properties required (see Fig. 4.43) but it is always between 180 and 650 °C. The temperature must be accurately controlled and hence the heating is done in an oil bath, salt bath, lead bath or air-circulating furnace.

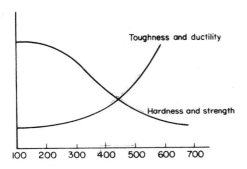

Fig. 4.43 How the properties of hardened steel are modified by tempering

Stabilising. This heat treatment is given to remove residual stresses which might cause distortion at a later stage.

1. *Stabilisation of steel.* To remove stresses after rough machining the component is heated at 525 to 550 °C for about 5 hours. To remove stresses after hardening and tempering the part is heated at about 150 °C for approximately 5 hours.

2. *Stabilisation of brass.* Cold worked brass sometimes suffers from season cracking in which cracks develop some time after working. To prevent this the component is given a stress relief treatment by heating it to 250 °C for about 1 hour.

3. *Stabilisation of aluminium alloys.* Castings, forgings and extrusions are often stabilised by heating to about 200 °C for about 5 hours, the duration depending upon the section thickness.

Exercise 4.2

1. Suggest suitable metals for each of the following: (a) a gas turbine blade, (b) an aircraft undercarriage, (c) a die cast lever, (d) a lathe bed, (e) a coupling for rolling stock, (f) a spanner, (g) a shaft carrying high stress.

2. Describe the nitriding process and state its advantages.

3. Describe the process of induction hardening. For which metals is it suitable?

4. State the main forms of supply for the following: (a) magnesium alloy, (b) aluminium bronze, (c) zinc alloy, (d) lead-antimony alloy.

5. What is a thermoplastic? What is a thermosetting plastic?

6. Suggest suitable plastics for each of the following: (a) sound insulation, (b) factory ducting, (c) a non-lubricated bearing, (d) a seal, (e) a tool handle, (f) a gear, (g) a foundry core.

7. Copy out the following table and for each metal write either poor (P), fair (F), good (G) or excellent (E) for each of the properties.

Material	Tensile strength	Hardness	Ductility	Machine-ability
Free cutting brass				
Phosphor bronze				
Austenitic chromium-nickel steel				
0·3% carbon steel				
High tensile steel				

8. Copy out the following table and for each of the plastics write either fair (F), good (G) or excellent (E) for each of the properties.

Material	Tensile strength	Compressive strength	Machineability
Rigid PVC			
PTFE			
Nylon			
Bakelite			
Glass fibre			

TABLE 1 PROPERTIES OF WROUGHT ALUMINIUM ALLOYS

Type	Composition %	Condition	0·1% P.S. MN/m²	U.T.S. MN/m²	Elong. %	Machine-ability	Cold forming
Non-heat treatable alloys	Aluminium 99·99%	Annealed ½ Hard Full hard	— — —	90 max. 100–120 130	30 8 5	Poor	Very good
	Copper 0·15 Silicon 0·6 Iron 0·7 Manganese 1·0 Zinc 0·1 Titanium 0·2 Aluminium 97·2	Annealed ¼ Hard ½ Hard ¾ Hard Full hard	— — — — —	115 max. 115–145 140–170 160–190 180	30 12 7 5 3	Fair	Very good
	Copper 0·1 Magnesium 7·0 Silicon 0·6 Iron 0·7 Manganese 0·5 Zinc 0·1 Chromium 0·5 Titanium 0·2 Aluminium 90·3	Annealed	90	310–360	18	Good	Fair
Heat treatable alloys	Copper 3·5–4·8 Magnesium 0·6 Silicon 1·5 Iron 1·0 Manganese 1·2 Titanium 0·3 Aluminium balance	Solution treated Fully heat treated	— —	380 420	— —	Good Very good	Good Poor
	Copper 0·1 Magnesium 0·4–1·5 Silicon 0·6–1·3 Iron 0·6 Manganese 0·6 Zinc 0·1 Chromium 0·5 Titanium 0·2 Aluminium balance	Solution treated Fully heat treated	110 230	185 280	18 10	Good Very good	Good Fair

TABLE 2   ALUMINIUM CASTING ALLOYS

Type	Composition %		Condition	0·2% P.S. MN/m²	U.T.S. MN/m²	Elong. %	Hardness BHN	Machineability
As cast	Copper 0·1 Magnesium 3–6 Silicon 10–13 Iron 0·6 Manganese 0·5 Nickel 0·1 Tin 0·05 Lead 0·1 Aluminium balance		Sand cast	60	160	5	50	Difficult
			Chill cast	70	190	7	55	Difficult
			Die cast	120	280	2	55	—
	Copper 0·7–2·5 Magnesium 0·3 Silicon 9·0–11·5 Iron 1·0 Manganese 0·5 Nickel 1·0 Zinc 1·2 Aluminium balance		Chill cast	100	180	1·5	85	Fair
			Die cast	150	320	1	85	—
Heat-treatable	Copper 4–5 Magnesium 0·1 Silicon 0·25 Iron 0·25 Manganese 0·1 Nickel 0·1 Zinc 0·1 Aluminium balance		Chill cast Fully heat treated	—	300	9	—	Good

The above alloys are used for food and chemical plant, motor car fittings, marine castings and hydraulic systems.

TABLE 3   PROPERTIES OF SOME COPPER ALLOYS

Name and uses	Composition %			Condition	Mechanical properties			
	Cu	Zn	Others		0·1% P.S. MN/m²	U.T.S. MN/m²	Elong. %	Vickers Hardness
Muntz metal for die stamping and extrusions	60	40	—	Extruded	110	350	40	75
Free cutting brass for high speed machining	58	39	Lead 3%	Extruded	140	440	30	100
Cartridge brass for severe cold working	70	30	—	Annealed	75	270	70	65
				Work hardened	500	600	5	180
Standard brass for press work	65	35	—	Annealed Work hardened	90 500	320 690	65 4	65 185
Admiralty gunmetal for general purpose castings	88	2	Tin 10%	Sand casting	120	290	16	85
Phosphor bronze for castings and bushes for bearings	Rem.		Tin 10% Phosphorus 0·03—0·25%	Sand casting	120	280	15	90

Cu = copper   Zn = zinc

TABLE 4   PROPERTIES OF CAST HIGH ALLOY STEELS

BS spec.	Type	Composition %							Mechanical properties			Applications etc.
		Cu	Si	Mn	Ni	Cr	Mo	C	U.T.S. MN/m²	Yield stress	Elong. %	
3100 BW10	Austenitic manganese steel	—	1·0	11·0	—	—	—	1·0				Possesses great hardness and hence is used for earth moving equipment, pinions, sprockets etc. where wear resistance is important.
3100 410 C21	13% chromium steel	—	1·0	1·0	1·0	13·5	—	0·15	540	370	15	Mildly corrosion resistant. Used in the paper industry.
3100 302 C25	Austenitic chromium-nickel steel	—	1·5	2·0	8·0	21·0	—	0·08	480	210	26	Cast stainless steel. Corrosion resistant and very ductile.
3100 315 C16	Austenitic chromium-nickel-molybdenum steel	—	1·5	2·0	10·0	20·0	1·0	0·08	480	210	22	Cast stainless steel with higher nickel content giving increased corrosion resistance. Molybdenum increases weldability.
3100 302 C25	Heat resisting alloy steel	—	2·0	2·0	10·0	22·0	1·5	0·4	560	3		Can withstand temperatures in excess of 650 °C.
3100 334 C11		—	3·0	2·0	65·0	10·0	1·0	0·75	460	3		Temperature at which scaling occurs is raised by increasing amount of chromium.

Yield stress in MN/m².
Cu = copper   Si = silicon   Mn = manganese   Ni = nickel   Cr = chromium   Mo = molybdenum   C = carbon

TABLE 5   PROPERTIES OF CARBON STEELS TO BS 970

Type	Composition %			Mechanical properties			Applications etc.
	C	Si	Mn	U.T.S. MN/m²	Elong. %	Hardness BHN	
070 M20	0·2	—	0·7	400	21	150	Easily machinable steels suitable for light stressing. Weldable.
070 M26	0·26	—	0·7	430	20	165	Stronger than En 2. Good machineability and is weldable.
080 M30	0·3	—	0·8	460	20	165	Increased carbon increases mechanical properties but slightly less machineable.
080 M36	0·36	—	0·8	490	18	180	Tough steel used for forgings, nuts and bolts, levers, spanners etc.
080 M40	0·4	—	0·8	510	16	180	Medium carbon steel which is readily machineable.
080 M46	0·46	—	0·8	540	14	205	Used for motor shafts, axles, brackets and couplings.
080 M50	0·5	—	0·8	570	14	205	Used where strength is more important than toughness, e.g. machine tool parts.
216 M28	0·28	0·25	1·3	540	10	180	Increased manganese content gives enhanced strength and toughness.
080 M15	0·15	0·25	0·8	460	16	——	Case hardening steel. Used where wear is important, e.g. gears, pawls, etc.

C = carbon   Si = silicon   Mn = manganese

TABLE 6   PROPERTIES OF MEDIUM AND LOW ALLOY STEELS

Type	Composition %								Mechanical properties		Applications etc.
	C	Si	Mn	Cr	Ni	Mo	W	V	U.T.S. MN/m²	Elong. %	
Low alloy structural steel	0·3	0·3	0·75	—	3	—			800	26	Crankshafts, high tensile shafts etc.
Nickel-chromium-molybdenum steel	0·35	0·3	0·7	0·8	2·8	0·7			1000	16	Air hardening steel. Used at high temperatures.
High tensile steel	0·4	—	—	1·2	1·5	0·3			1800	14	Used where high strength is needed.
Spring steel	0·5	1·6	1·3						1500		
Steel for cutting tools	1·2	—	—	1·5	—	—	4	0·3			
Die steel	0·35	—	0·3	5·0	—	1·4		0·4			

C = carbon   Si = silicon   Mn = manganese   Cr = chromium
Ni = nickel   Mo = molybdenum   W = tungsten   V = vanadium

TABLE 7 PROPERTIES OF STAINLESS STEELS

BS ref.	Type	Composition %			Mechanical properties				Applications etc
		C	Cr	Others	0·2% P.S. MN/m²	U.T.S. MN/m²	Elong. %	Hardness Vickers	
410 S21	Martensitic stainless steel	0·12	13	—	420	590	20	170	Not suitable for welding or cold forming. Possesses moderate machine-ability. Used for applications where resistance to tempering at high temperature is important e.g. turbine blades.
431 S29		0·15	16	2·5%Ni	740	900	11	270	
430 S15	Ferritic stainless steel	0·06	16	—	370	540	20	165	More corrosion resistant than the martensitic steels. They are hardenable by heat treatment. Used for press work because of high ductility.
302 S25	Austenitic stainless steel	0·08	18	9·0% Ni	210	510	40	170	Possesses good resistance to corrosion, good weldability, toughness at low temperature and excellent ductility. May be hardened by cold working.

C = carbon   Cr = chromium   Ni = nickel

TABLE 8   PROPERTIES OF HIGH TENSILE STEELS

BS ref.	Type	Composition %								Mechanical properties			Applications etc.
		C	Si	Mn	Ni	Cr	Mo	Co	Ti	U.T.S. MN/m²	0·2 P.S. MN/m²	Elong. %	
817 M40	Direct hardening nickel steel	0·44	0·35	0·7	1·7	1·4	0·35	—	—	1540	1240	8	Used where weight saving is important for instance in the aircraft industry. The deep hardening types are used for plastic moulding dies, shear blades, cold drawing mandrels and pressure vessels. These steels are all difficult to machine.
	Direct hardening chrome-molybdenum steel	0·35	0·35	0·65	0·4	3·5	0·7	—	—	1540	1240	10	
	Maraging steels	—	—	—	18	—	3·0	8·5	0·20	1480	1400	14	

C = carbon   Si = silicon   Mn = manganese   Ni = nickel
Cr = chromium   Mo = molybdenum   Co = cobalt   Ti = titanium

TABLE 9   PROPERTIES OF SOME THERMOPLASTICS

Name	Uses etc.	Tensile strength	Compressive strength	Machineability
PVC (rigid)	Tough and abrasive resistant. Used for insulation of electric cables, factory ducting, radio and TV parts	E	G	E
Polystyrene	Crystal clear and resistant to moisture. Bathroom fittings, sound insulation and aircraft panels.	E	G	F
PTFE	Tough and flexible and low coefficient of friction. Gaskets, filters, non-lubricating bearings, pipes and seals.	F	G	E
Polypropylene	Can withstand high temperatures and is resistant to most liquids. Pipes and pipe fittings, tool handles and heater ducts.	F	F	E
Nylon	High melting point, wear resistant and low coefficient of friction. Used for gears, bushes, bearings, flexible couplings and cable sheathing.	E	G	E

F = Fair (25 MN/m²) } Tensile
E = Excellent (55 MN/m²) } strength

F = Fair (40 MN/m²) } Compressive
G = Good (180 MN/m²) } strength

TABLE 10 PROPERTIES OF SOME THERMOSETTING PLASTICS

Name	Uses etc.	Tensile strength	Compressive strength	Machineability
Epoxy resin (glass fibre filled)	Laminates for circuit boards, circuit breaker equipment etc.	O	E	G
Melamine formaldehyde (cellulose filled)	Light (slightly heavier than wood) and durable, hard and resistant to abrasion. Good electrical properties. Sockets, switches and foundry cores.	G	G	F
Bakelite	Rigid and a good insulator. Used for electrical plugs and switches, radio cabinets etc.	G	G	F
Polyester (glass fibre filled)	Hard and tough and good electrical insulators. Used for motor car bodies, aircraft parts etc. This is the well known glass fibre.	E	G	G
Silicone (asbestos filled)	Used for high temperature electrical insulation, etc. Obtainable as sheet, tubes etc.	O	G	F

O = outstanding (80 MN/m²)
E = excellent (55 MN/m²) } Tensile strenth
G = good (40 MN/m²)

E = excellent (250 MN/m²) } Compressive strength
G = good (180 MN/m²)

Part 5   Drawing

Geometry of screw threads (Fig. 5.1).

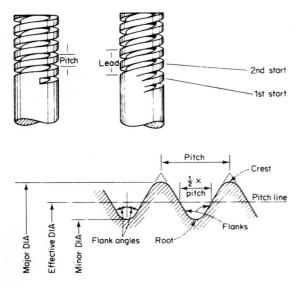

Fig. 5.1   The names given to the various parts of a screw thread

1. The *pitch* of a thread is the distance between corresponding points on adjacent threads measured parallel to the axis of the thread.
2. The *lead* of a thread is the distance that the thread moves in one revolution.
3. A *single-start thread* has the lead equal to the pitch.
4. A *multi-start thread* has a number of separate threads which are formed simultaneously side by side. The lead of a multi-start thread is pitch × number of starts.
5. The *major* or *outside diameter* is the greatest diameter of the thread.
6. The *minor* or *root diameter* is the smallest diameter of the thread.
7. The *effective* or *pitch diameter* is the diameter between lines drawn through the thread profile on both sides of the screw where the width of the groove equals half the pitch.

Thread forms (Fig. 5.2).

1. *British Standard Whitworth* (BS 84). This is a symmetrical V-thread with an angle of 55° between the flanks.
2. *British Standard Fine* (BS 84). This is a thread of Whitworth form but for corresponding diameters the pitches are finer and hence it is used for screws subject to vibration. Fig. 5.3 gives a comparison between a $\frac{1}{2}$ in diameter BSW and a $\frac{1}{2}$ in diameter BSF thread.
3. *British Association* (BS 93). This is a symmetrical V-thread with an angle of $47\frac{1}{2}°$ between the flanks. It is used for small diameter work because the larger root radius and slightly smaller depth of thread makes it easier to produce than BSW for screws below $\frac{1}{4}$ in diameter. The BA threads are numbered 0 BA to 25 BA, 0 BA being the largest and 25 BA the smallest. Like the BSW and BSF threads the BA thread is being replaced by similar sizes in ISO inch and ISO metric threads.
4. *ISO unified* (BS 1580). This is a symmetrical V-thread with an angle of 60° between the flanks. The unified thread system is a constant pitch series and it has a fixed number of threads per inch i.e. 4, 6, 8, 12, 16, 28, 32. They are designated on drawings by their major diameters and the number of threads per inch followed by the type and class of the thread. There are three classes of bolts and nuts:
Class 1A bolt, Class 1B nut—commercial quality
Class 2A bolt, Class 2B nut—better grade of thread
Class 3A bolt, Class 3B nut—finest quality thread
There are three types of thread: coarse (UNC), fine (UNF) and extra fine (UNEF). Some examples are $\frac{1}{4}$"/20/UNF/3A—$\frac{1}{4}$" dia × 20 tpi fine bolt thread of finest quality.
$\frac{5}{8}$"/11/UNC/1B—$\frac{5}{8}$" dia × 11 tpi coarse nut thread of commercial quality.
5. *ISO metric* (BS 3643). The thread form is identical with the unified thread and like the unified thread is a constant pitch series. There are two series of diameters with graded pitches, one with coarse threads and the other with fine threads. There are three classes of bolts and nuts:
Class 4h bolt—5H nut—for precision threads
Class 6g bolt—6H nut—for commercial quality threads
Class 8g bolt—8H nut—for coarse quality threads.
Coarse threads are marked on drawings by the symbol M followed by the thread diameter and the class. Thus

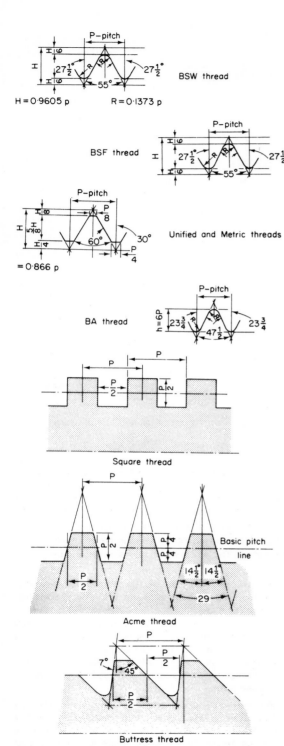

Fig. 5.2   Forms of various types of thread

M 20—6g—20 mm diameter coarse bolt thread of commercial quality. Fine threads are marked with the symbol M, the diameter and a multiplication sign followed by the pitch in millimetres and the class. Thus

M 30×2—6 H—30 mm diameter fine nut thread having a pitch of 2 mm of commercial quality.

6. *Square thread.* This is used in machine tools and power transmitting mechanisms. However square threads are difficult to produce and because adjustment is impossible backlash increases with wear. Hence for most power transmitting mechanisms the square thread has been replaced by the acme thread.

7. *Acme thread* (BS 1104). This is a symmetrical V-thread with an angle of 29° between the flanks. The depth of thread is half the pitch.

8. *Buttress thread* (BS 1657). This thread is used when heavy loads in one direction only are to be resisted as occurs in presses and vices.

Fig. 5.3   Comparison of ½ in BSW and ½ in BSF threads

Interchangeability. This occurs when one part can be substituted for a similar part which has been made to the same drawing. For the interchangeability of holes and shafts the ISO system of limits and fits (BS 4500) is used.

ISO system of limits and fits. The limits on a shaft may be specified by means of a deviation from a basic size and a tolerance (Fig. 5.4). By varying the deviation various kinds of fit are obtained (Fig. 5.5). The ISO system provides 28 different deviations which allow for every conceivable type of fit. The deviations for the holes

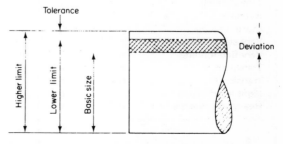

Fig. 5.4   Limits specified by means of a deviation from the basic size and a tolerance.

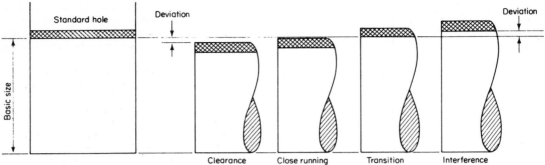

Fig. 5.5   The deviation determines the type of fit

are denoted by the capital letters A, B, C, CD, D, E, EF, F, FG, GH, JS, J, K, M, N, P, R, S, T, U, V, X, Y, Z, ZA, ZB and ZC. Holes A to G are larger than the basic size whilst holes J to ZC are smaller than the basic size. The deviations on the shafts are denoted by the small letters a, b, c etc. shafts a to g are smaller than the basic size whilst shafts j to zc are larger than the basic size.

The amount of tolerance needed depends upon the quality of the work. Sand castings for instance require very large tolerances whilst gauges require very small tolerances. To cover the vast range of work needed in engineering 18 grades of tolerance numbered 01, 0, 1, 2, 3, 4, ... 15, 16 are used.

A hole is designated, for example, by the symbols H 7 and a shaft by the symbols d 10.

Whilst BS 4500 provides for a great many hole and shaft tolerances, experience has shown that only a few are generally needed in engineering. The table below gives details of selected fits which will meet most engineering requirements.

Type of fit	Hole	Shaft
Loose running	H 11	c 11
Easy running	H 9	d 10
Running	H 9	e 9
Close running	H 8	f 7
Very close running	H 7	g 6
Very fine running	H 7	h 6
Transition	H 7	k 6
Transition	H 7	n 6
Light interference	H 7	p 6
Heavy interference	H 7	s 6

Fig. 5.6 shows how the system is used on drawings and Table 11 (p. 124) gives details of the holes and shafts.

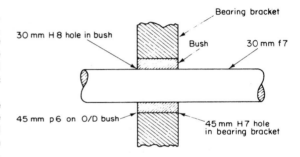

Fig. 5.6   The use of BS 4500 on drawings

Geometrical tolerances (BS 308: Part 3: 1972). All manufactured parts will have errors in their shape. These errors are called geometrical errors and in order to control them geometrical tolerances are placed on the drawing.

Tolerance symbols (Fig. 5.7). BS 308 recommends the use of symbols to denote the various types of geometrical tolerances. The way in which they are used is shown in Fig. 5.8.

The tolerance frame (Fig. 5.9). The geometrical tolerance is indicated in a rectangular frame which is divided into compartments as shown in the diagram. The symbol for the characteristic (see Fig. 5.7) is placed in the left-hand compartment and the tolerance value is shown in the next compartment. If a datum has to be specified, a third compartment is used in which is placed the datum identification letter. Examples of the use of datum letters are given later.

Tolerances applicable to restricted lengths of feature. If a tolerance is applied only to a particular part of a component, it is indicated as shown in Fig. 5.10. If the tolerance applies to a specific length it is indicated as shown in Fig. 5.11.

	Type of tolerance	Characteristic to be toleranced	Symbol
For single features	Form	Straightness	—
		Flatness	⟋⟋
		Roundness	○
		Cylindricity	⌀
		Profile of a line	⌒
		Profile of a surface	⌓
For related features	Attitude	Parallelism	//
		Squareness	⊥
		Angularity	∠
	Location	Position	⊕
		Concentricity	◎
		Symmetry	⩵
	Composite	Run-out	↗

Fig. 5.7   Tolerance symbols

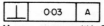

Means squareness within 0·03 mm
with reference to datum A

Fig. 5.9   The tolerance frame

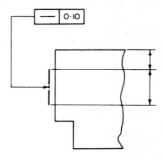

Fig. 5.10

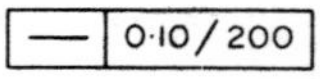

—	0·10 / 200

Fig. 5.11

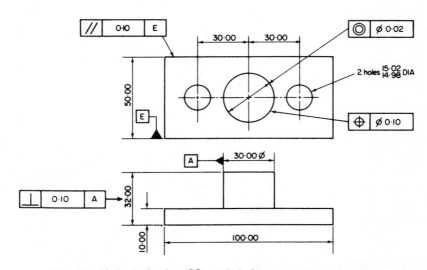

Fig. 5.8   Method of using ISO symbols for geometrical tolerances

The maximum material condition. The worst condition for the assembly of parts occurs when the mating parts are at their maximum material limits and there are present geometrical errors also at their maximum. This means that the size of a hole is reduced to its minimum and the size of a shaft is increased to its maximum. To allow for this the symbol Ⓜ is used as shown in Fig. 5.12. It means that the axis of the detail is to be contained in a cylinder whose diameter is 0·05 mm, the diameter of the part being at its maximum material limit. (see discussion below on straightness).

Dimensions defining true positions. Dimensions which define true positions are enclosed in a box thus 50 or ⌀ 30 (see Fig. 5.17).

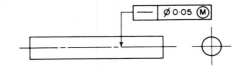

Fig. 5.12

Tolerance for straightness (Fig. 5.13). This tolerance is used to limit the straightness of a line on a surface or the straightness of an axis.

Scribed line

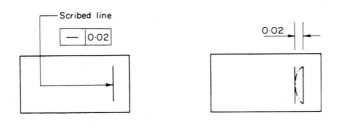

The line shown on the surface has to lie between two parallel straight lines on the surface 0·02 mm apart.

(a)

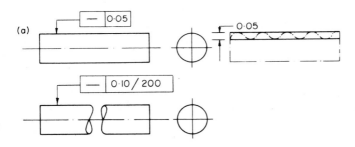

The surface of the feature has to lie between two parallel straight lines 0·05 apart lying in an axial plane.

As above but any portion of length 200 mm is to be taken.

(b)

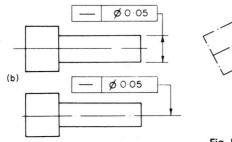

The axis of that part of the detail marked is to be contained in a cylinder whose diameter is 0·05 mm.

As above, but the tolerance stated applies to the whole piece.

Fig. 5.13 Straightness tolerance

Tolerance for flatness (Fig. 5.14). Flatness is toleranced by two perfectly flat parallel planes, the tolerance value being the distance between the planes.

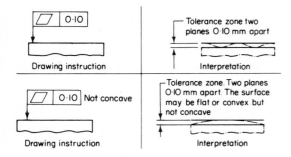

Fig. 5.14   Flatness tolerance

Tolerance for roundness (Fig. 5.15). A roundness tolerance controls the errors in form of a circle in the plane in which it lies. In the case of a solid of revolution, such as a cylinder or a cone, the tolerance controls the form of the circle in a plane perpendicular to the axis. For a sphere it controls the form of the maximum diameter.

Tolerance for cylindricity (Fig. 5.15). A shaft will have errors of cylindricity while a disc will have only errors of roundness. Errors of roundness may occur as ovality or lobing (Fig. 5.16), but errors in cylindricity may occur as errors in roundness, straightness and parallelism.

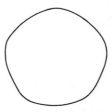

Fig. 5.16   Lobed figure with a constant "diameter"

Profile tolerance of a line. The theoretical shape of a profile is defined by boxed dimensions as shown in Fig. 5.17. The tolerance zone is then specified in relation to the theoretical profile. A bilateral tolerance equally disposed about the theoretical profile is intended unless the drawing indicates a unilateral tolerance as shown in profile 2 of Fig. 5.17.

Profile tolerance of a surface. The theoretical form of the surface is defined by boxed dimensions and the tolerance zone is established in relation to this theoretical form. A typical example is shown in Fig. 5.18.

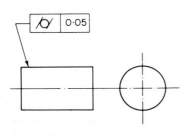

The curved surface has to lie between two cylindrical surfaces coaxial with each other and a radial distance of 0·05 mm apart.

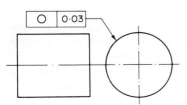

The periphery at any cross-section perpendicular to the axis has to lie between two circles concentric with each other and a radial distance of 0·03 mm apart, in the plane of the section.

Fig. 5.15   Tolerances of roundness and cylindricity

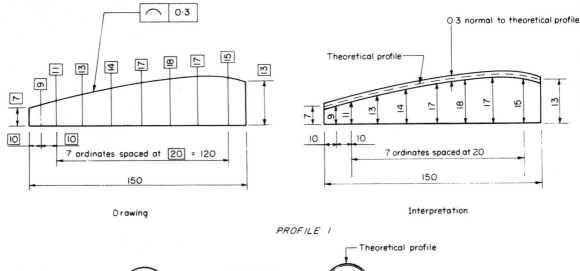

Drawing

Interpretation

PROFILE 1

Drawing

Interpretation

PROFILE 2

Fig. 5.17

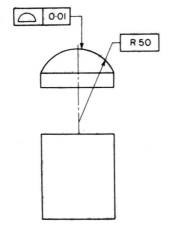

Fig. 5.18

Tolerance for parallelism (Fig. 5.19). Parallelism involves two or more features and one of these must be taken as the datum. The datum is indicated by a letter which appears in the tolerance frame.

Tolerance of squareness (Fig. 5.20). It is necessary to refer to a datum in order to specify the squareness of a surface with a datum line or a datum surface.

Tolerance for angularity (Fig. 5.21). As with squareness, a datum is required to give an angular tolerance.

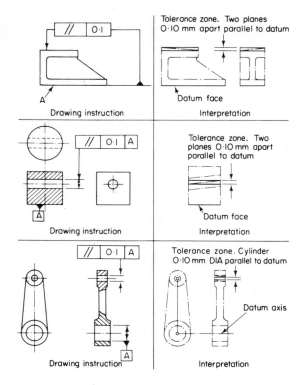

Fig. 5.19 Parallelism tolerance

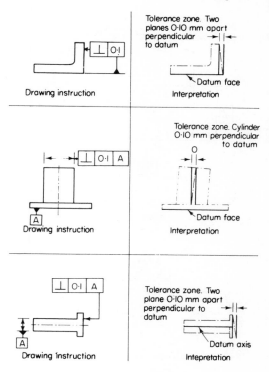

Fig. 5.20 Squareness tolerance

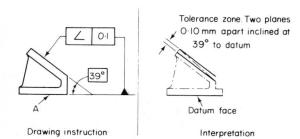

Fig. 5.21 Angularity tolerance

Tolerance of location. These control the position of the feature and may also control its form. The chief tolerances are as follows:

(1) *Tolerances of position* which limit the deviation of the position of a feature from its true position (Fig. 5.22).

(2) *Tolerances of concentricity* (Fig. 5.23). Tolerances limiting concentricity either control the position of one feature with respect to another or they restrain two or more features by specifying a common zone.

(3) *Tolerances of symmetry* (Fig. 5.24). These are given with reference to a datum, often a centre-line as shown in the diagrams.

Tolerances of run out (Fig. 5.25). The run-out tolerance represents the maximum permissible variation of a fixed point during one complete revolution about the datum axis. As shown in the diagrams, it is measured by means of a dial indicator.

This is only a brief summary of the main geometrical tolerances. For a full account you are recommended to consult the standard BS 308:Part 3.

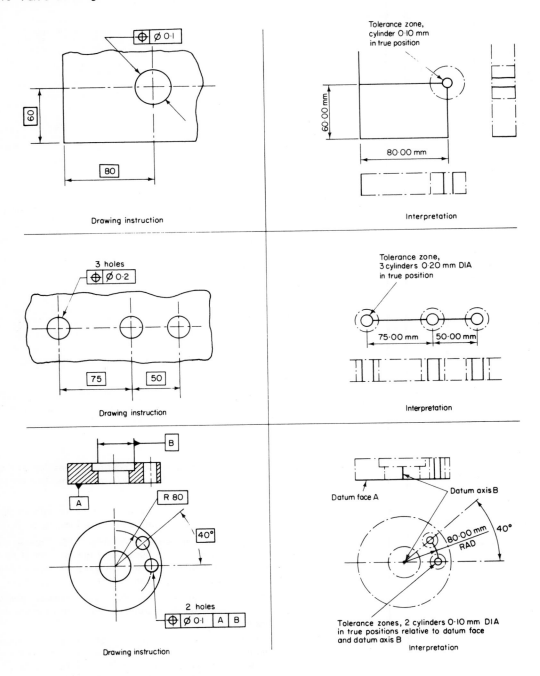

Fig. 5.22   Positional tolerances

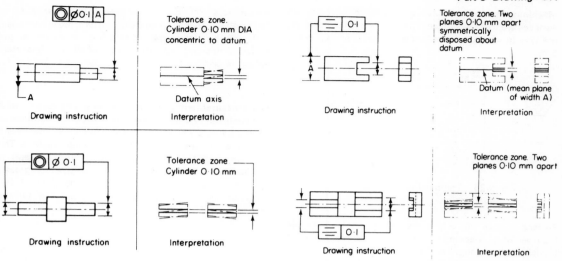

Fig. 5.23 Concentricity tolerances

Fig. 5.24 Symmetry tolerance

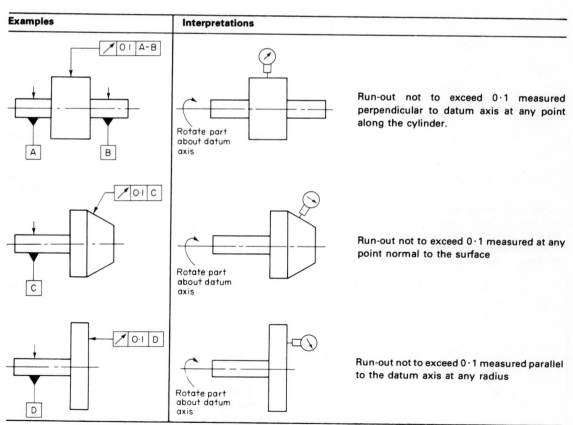

Fig. 5.25 Tolerances of run out

Machining symbols. These are used to indicate that surfaces are to be machined. The general symbol used and the way in which it is used on drawings is shown in Fig. 5.26.

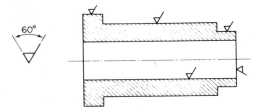

Fig. 5.26　The machining symbol

Surface texture. Fig. 5.27 shows, exaggerated, the main features of a machined surface. The primary texture (usually called *roughness*) is of short wavelength. The secondary texture (or *waviness*) is of much longer wavelength. The *lay* is the directional nature of the texture pattern.

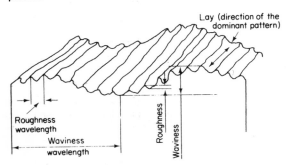

Fig. 5.27　The characteristics of a machined surface

The roughness is due to the shape of the tool, the feed rate and tool chatter. The waviness is due to such things as the misalignment of centres, inconsistent feed motion etc.

Fig. 5.28 shows, on a magnified scale, the cross-sections of three surfaces. In each case the departure from

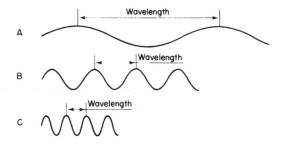

Fig. 5.28　Wavelength is important to surface texture

smoothness has been represented by a wave form. Although the peak to valley height is the same in each case, surface A is usually regarded as being the smoothest and surface B the roughest. Hence wavelength is an important aspect of surface finish.

The measurement of surface texture. The quality of a surface finish is usually measured by special instruments having a very sensitive stylus which follows the peaks and valleys as it moves across the machined surface. The movement of the stylus is recorded and a graph is produced by the instrument which gives a greatly magnified illustration of surface roughness (Fig. 5.29). Most of the long wave effects are nullified by the instrument.

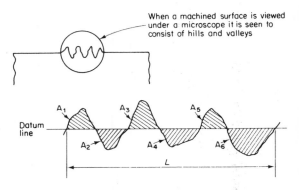

Fig. 5.29　Magnified illustration of surface roughness

Centre-line average (C.L.A.) method. When a graph such as that shown in Fig. 5.29 is produced, a centre-line is found such that the areas above the datum line are equal to the areas below. Thus in Fig. 5.29

$$A_1 + A_3 + A_5 = A_2 + A_4 + A_6$$

The C.L.A. value is then given by:

C.L.A. number

$$= \frac{\text{the sum of the areas above and below the datum line}}{\text{the sampling length}}$$

$$= \frac{A_1 + A_2 + A_3 + A_4 + A_5 + A_6}{L}$$

The C.L.A. number is always stated in micrometres (μm).

Surface texture symbol. Fig. 5.30 shows the methods used to indicate, on drawings, the quality of the surface finish. When it is necessary to specify the direction of lay it is indicated as shown in Fig. 5.31.

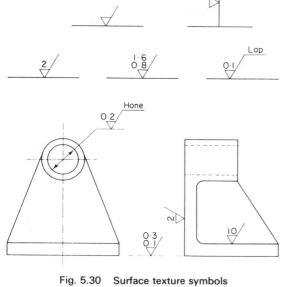

Fig. 5.30   Surface texture symbols

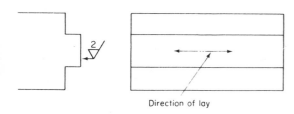

Fig. 5.31   Specifying the direction of lay

Welding symbols (BS 499 Part 2). On a drawing it is necessary to state the type of weld required and its position. This is done by (*a*) a weld symbol (Fig. 5.32), (*b*) an arrow and a reference line (Fig. 5.33).

When the weld is required on the arrow side of the joint (see Figs. 5.34 and 5.35) the welding symbol is placed *under* the reference line. If the weld is required on the other side of the joint the weld symbol is placed on the top side of the reference line.

Fig. 5.34 shows all the possible ways of representing a single fillet weld whilst Fig. 5.35 shows fillet welds on each side of the joint. Fig. 5.36 shows the representation of single V butt welds.

Form of weld	Sectional representation	Appropriate symbol
Fillet	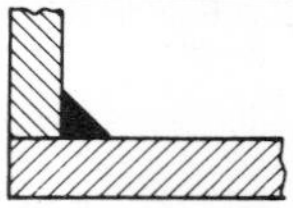	
Square butt		
Single-V butt		
Double-V butt		
Single-U butt		
Double-U butt	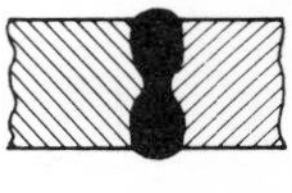	
Single-bevel butt	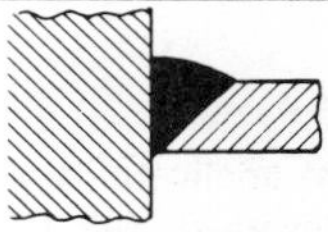	
Double-bevel butt	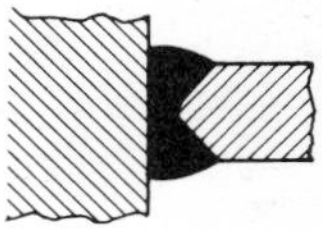	
Single-J butt	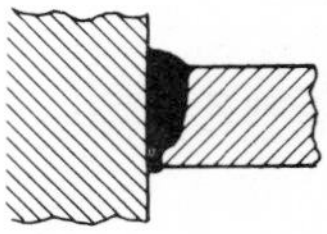	
Double-J butt	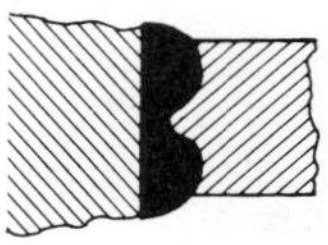	

Fig. 5.32   Welding symbols

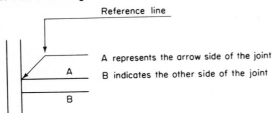

A represents the arrow side of the joint

B indicates the other side of the joint

Fig. 5.33

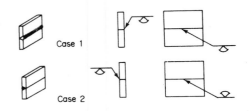

Fig. 5.36   Single-V butt welds (without sealing run)

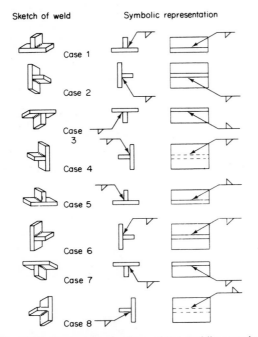

Fig. 5.34 Typical applications of the welding symbols for fillet welds (all in third-angle projection)

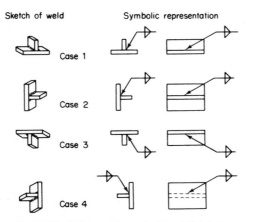

Fig. 5.35   Fillet welds each side of T-joint

Electrical symbols (BS 3939). The standard BS 3939 shows symbols used to represent practically every electrical and electronic device. Some of the more widely used symbols are shown in Fig. 5.37.

Description	Symbol	Description	Symbol
Switch		Link with 2 separable contacts	
Single cell		Link with 2 bolted contacts	
Battery of cells		Plug (male)	
Earth		Socket (female)	
Terminal or tag	o	Fuse	or
Fixed resistor		Variable resistor	
Resistor with moving contact		Cable with 4 conductors	
Generator	G	Motor	M
Isolator		Ammeter	A
Voltmeter	V	Filament lamp	

Fig. 5.37   Electrical symbols

Some examples of electrical circuit diagrams are given in Fig. 5.38. Note that switches, fuses etc. are always wired to the live conductor. It is dangerous to wire switches to the neutral conductor because current may flow even when the device is in the 'off' position.

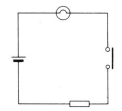

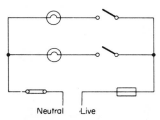

Circuit containing a battery, lamp, switch and resistor.

A lighting circuit with two switches.

Neutral   Live

Fig. 5.38   Electrical circuit diagrams

Exercise 5.1

1. Draw 5 times full size the following threads:
   (a) BSW thread 1″ diameter × 8 tpi
   (b) BSF thread 1″ diameter × 10 tpi
   (c) Metric thread M 20 × 1·5.
2. Explain, using diagrams, the following terms: (a) lead of a thread (b) pitch of a thread (c) root (d) flank (e) major diameter (f) minor diameter (g) effective diameter.
3. The following are examples of Metric bolt threads as called up on a drawing. For each example state:
   (i)   The type of thread i.e. coarse or fine.
   (ii) The major diameter of the thread.
   (iii) Whether the thread is precision, commercial or coarse grade.
   (a) M 30 − 6g   (b) M 24 × 1·5 − 4h
   (c) M 20 × 1·5 − 8g   (b) M 8 − 4h.
4. Write down how the following metric threads would be called up on a drawing: (a) coarse nut thread 20 mm diameter, commercial quality (b) fine nut thread 36 mm diameter × 3 mm pitch, precision grade (c) coarse bolt thread 30 mm diameter, coarse grade.
5. Sketch (a) an acme thread (b) a buttress thread. Why is an acme thread often used instead of a square thread? When is a buttress thread used?
6. Using Table 11 (p. 124) state the limits of size for the shaft and the hole for each of the following:
   (a) 240 mm diameter to give a light interference fit.

(b) 110 mm diameter to give a transition fit.
(c) 300 mm diameter to give a close running fit.
(d) 350 mm diameter to give an easy running fit.
In each case state the maximum amount of clearance or interference.

7. A gunmetal bush is to be fitted into a 30 mm nominal bore hole in a 20 mm thick plate. The bush is of 20 mm nominal bore to receive the machined end of a layshaft. The allowance is to be at least 0·020 mm and the maximum clearance not more than 0·062 mm. Using Table 11 (p. 124):
   (a) Draw a fully dimensional sketch of the bush twice full size.
   (b) State the tolerance grade on the shaft.
   (c) State the tolerance grade on the hole in the plate.
   (d) Find the maximum interference between the bush and the hole.
8. A crankshaft and main bearing assembly is shown in Fig. 5.39:
   (a) State the maximum and minimum clearance or interference that is provided by the symbols shown using Table 11 (p. 124).
   (b) Sketch a GO—NOT GO plug gauge that could be used to check the bore of the bush giving the diameter at each end.

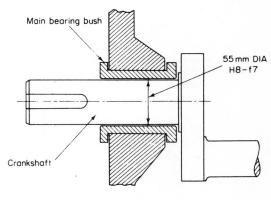

Main bearing bush

55 mm DIA
H8−f7

Crankshaft

Fig. 5.39

9. Fig. 5.40 shows three surface texture symbols. Explain the meaning of each.

0·8   2·0   2·0 Grind
           1·2

Fig. 5.40

10. What is meant by the abbreviation TP? Show on sketches the two methods used for tolerancing a profile.

11. Fig. 5.41 shows the drawing requirement and the actual dimensions obtained by measuring. State which are outside limits and state the amount of error.

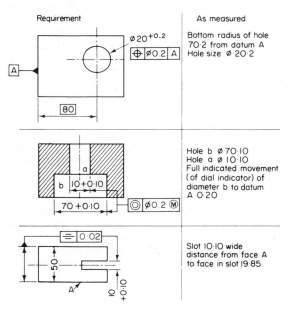

Requirement	As measured
	Bottom radius of hole 70·2 from datum A Hole size ⌀ 20·2
	Hole b ⌀ 70·10 Hole a ⌀ 10·10 Full indicated movement (of dial indicator) of diameter b to datum A 0·20
	Slot 10·10 wide distance from face A to face in slot 19·85.

Fig. 5.41

12. Fig. 5.42 shows a detail in which the ISO symbols for geometric characteristics have been used. For each state the meaning.

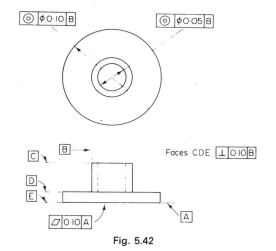

Fig. 5.42

13. Interpret the welding instructions given in Fig. 5.43.
14. Fig. 5.44 shows several welded assemblies. Using the standard welding symbols make a drawing of each assembly.

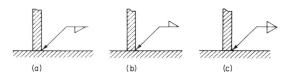

Fig. 5.43

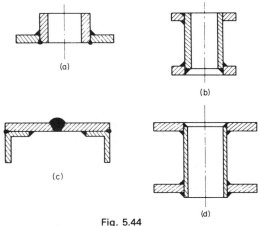

Fig. 5.44

15. Fig. 5.45 shows some electrical circuit diagrams. Name the parts marked with the capital letters.

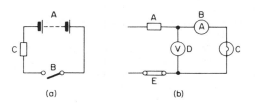

Fig. 5.45

16. Each component in · the following list can be represented by one of the symbols shown in Fig. 5.46. State which. Voltmeter, switch, battery of many cells, light bulb, fixed resistor, fuse, variable resistor, single cell, ammeter, isolator, earth, plug, socket.
17. Draw circuit diagrams for the following:
    (a) two lamps in series controlled by one switch
    (b) two lamps in parallel each controlled by separate switches.

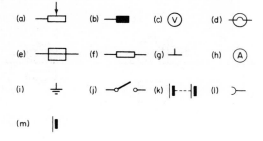

Fig. 5.46

TABLE 11 SELECTION OF PRIMARY FITS

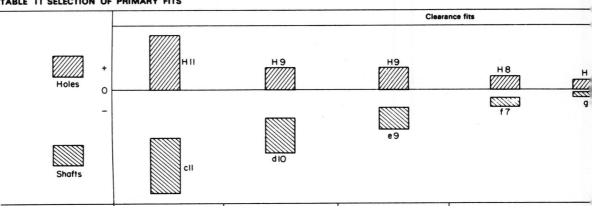

Nominal sizes		Tolerance		Tolerance		Tolerance		Tolerance		
Over	To	H11	c11	H9	d10	H9	e9	H8	f7	H
mm	mm	0·001 mm	0·001 mm	0·001 mm	0·001 mm	0·001 mm	0·001 mm	0·001 mm	0·001 mm	0·001
—	3	+60 / 0	−60 / −120	+25 / 0	−20 / −60	+25 / 0	−14 / −39	+14 / 0	−6 / −16	+1 / 0
3	6	+75 / 0	−70 / −145	+30 / 0	−30 / −78	+30 / 0	−20 / −50	+18 / 0	−10 / −22	+1 / 0
6	10	+90 / 0	−80 / −170	+36 / 0	−40 / −98	+36 / 0	−25 / −61	+22 / 0	−13 / −28	+1 / 0
10	18	+110 / 0	−95 / −205	+43 / 0	−50 / −120	+43 / 0	−32 / −75	+27 / 0	−16 / −34	+1 / 0
18	30	+130 / 0	−110 / −240	+52 / 0	−65 / −149	+52 / 0	−40 / −92	+33 / 0	−20 / −41	+2 / 0
30	40	+160 / 0	−120 / −280	+62 / 0	−80 / −180	+62 / 0	−50 / −112	+39 / 0	−25 / −50	+2 / 0
40	50	+160 / 0	−130 / −290							
50	65	+190 / 0	−140 / −330	+74 / 0	−100 / −220	+74 / 0	−60 / −134	+46 / 0	−30 / −60	+3 / 0
65	80	+190 / 0	−150 / −340							
80	100	+220 / 0	−170 / −390	+87 / 0	−120 / −260	+87 / 0	−72 / −159	+54 / 0	−36 / −71	+3 / 0
100	120	+220 / 0	−180 / −400							
120	140	+250 / 0	−200 / −450							
140	160	+250 / 0	−210 / −460	+100 / 0	−145 / −305	+100 / 0	−84 / −185	+63 / 0	−43 / −83	+4
160	180	+250 / 0	−230 / −480							
180	200	+290 / 0	−240 / −530							
200	225	+290 / 0	−260 / −550	+115 / 0	−170 / −355	+115 / 0	−100 / −215	+72 / 0	−50 / −96	+4
225	250	+290 / 0	−280 / −570							
250	280	+320 / 0	−300 / −620	+130 / 0	−190 / −400	+130 / 0	−110 / −240	+81 / 0	−56 / −108	+5
280	315	+320 / 0	−330 / −650							
315	355	+360 / 0	−360 / −720	+140 / 0	−210 / −440	+140 / 0	−125 / −265	+89 / 0	−62 / −119	+5 / 0
355	400	+360 / 0	−400 / −760							
400	450	+400 / 0	−440 / −840	+155 / 0	−230 / −480	+155 / 0	−135 / −290	+97 / 0	−68 / −131	+6
450	500	+400 / 0	−480 / −880							

			Transition fits				Interference fits			

	Tolerance		Tolerance		Tolerance		Tolerance		Tolerance	
g6	H7	h6	H7	k6	H7	n6	H7	p6	H7	s6
0·001 mm	0·001 mm	0·001 mm	0·001 mm	0·001 mm	0·001 mm	0·001 mm	0·001 mm	0·001 mm	0·001 mm	0·001 mm
−2 / −8	+10 / 0	−6 / 0	+10 / 0	+6 / +0	+10 / 0	+10 / +4	+10 / 0	+12 / +6	+10 / 0	+20 / +14
−4 / −12	+12 / 0	−8 / 0	+12 / 0	+9 / +1	+12 / 0	+16 / +8	+12 / 0	+20 / +12	+12 / 0	+27 / +19
−5 / −14	+15 / 0	−9 / 0	+15 / 0	+10 / +1	+15 / 0	+19 / +10	+15 / 0	+24 / +15	+15 / 0	+32 / +23
−6 / −17	+18 / 0	−11 / 0	+18 / 0	+12 / +1	+18 / 0	+23 / +12	+18 / 0	+29 / +18	+18 / 0	+39 / +28
−7 / −20	+21 / 0	−13 / 0	+21 / 0	+15 / +2	+21 / 0	+28 / +15	+21 / 0	+35 / +22	+21 / 0	+48 / +35
−9 / −25	+25 / 0	−16 / 0	+25 / 0	+18 / +2	+25 / 0	+33 / +17	+25 / 0	+42 / +26	+25 / 0	+59 / +43
−10 / −29	+30 / 0	−19 / 0	+30 / 0	+21 / +2	+30 / 0	+39 / +20	+30 / 0	+51 / +32	+30 / 0	+72 / +53
									+30 / 0	+78 / +59
−12 / −34	+35 / 0	−22 / 0	+35 / 0	+25 / +3	+35 / 0	+45 / +23	+35 / 0	+59 / +37	+35 / 0	+93 / +71
									+35 / 0	+101 / +79
−14 / −39	+40 / 0	−25 / 0	+40 / 0	+28 / +3	+40 / 0	+52 / +27	+40 / 0	+68 / +43	+40 / 0	+117 / +92
									+40 / 0	+125 / +100
									+40 / 0	+133 / +108
−15 / −44	+46 / 0	−29 / 0	+46 / 0	+33 / +4	+46 / 0	+60 / +31	+46 / 0	+79 / +50	+46 / 0	+151 / +122
									+46 / 0	+159 / +130
									+46 / 0	+169 / +140
−17 / −49	+52 / 0	−32 / 0	+52 / 0	+36 / +4	+52 / 0	+66 / +34	+52 / 0	+88 / +56	+52 / 0	+190 / +158
									+52 / 0	+202 / +170
−18 / −54	+57 / 0	−36 / 0	+57 / 0	+40 / +4	+57 / 0	+73 / +37	+57 / 0	+98 / +62	+57 / 0	+226 / +190
									+57 / 0	+244 / +208
−20 / −60	+63 / 0	−40 / 0	+63 / 0	+45 / +5	+63 / 0	+80 / +40	+63 / 0	+108 / +68	+63 / 0	+272 / +232
									+63 / 0	+292 / +252

Answers to exercises

Exercise 1.1
1. (a) 362·1
   (e) 0·000 4648
   (i) 71·93
   (b) 17 970
   (f) 2·258
   (j) 0·1071
   (c) 148 900
   (g) 2·599
   (k) 0·1579
   (d) 0·3561
   (h) 0·1009
   (l) 0·000 006 673

Exercise 1.2
1. 863·2
5. 0·000 5508
2. 393·2
3. 0·4837
4. 0·00 1306

Exercise 1.3
1. 2·486
5. 0·0 9210
2. 1·067
3. 0·5069
4. 0·4119

Exercise 1.4
1. 102·0
5. 0·000 0044
9. 0·5338
2. 6·832
6. 34 400
10. 1·762
3. 48·16
7. 16·95
4. 1·212
8. 1·883

Exercise 1.5
1. 2·792
5. 0·070 52
9. 1·621
2. 15·37
6. 1·639
10. 0·014 89
3. 0·9210
7. 0·4891
4. 8·504
8. 0·5230

Exercise 1.6
1. 3·125 kW
5. 6·073 kW
9. 14·73 m/min
2. 0·374
6. 0·0147
10. 99·01 μm
3. 44·18 cm³/min
7. 4·169
11. 9·511 μm
4. 20·63 m/min
8. 58·88 Nm

Exercise 1.7
1. 47·11
2. 50·83
3. 9·166
4. 1193
5. (a) 21·14 (b) 0·2320 (c) 0·223 (d) 0·346 (e) 15·17 (f) 0·0654 (g) 0·2418 (h) 0·0592

Exercise 1.8
1. $\dfrac{C}{\pi}$
2. $\dfrac{S}{\pi d}$
3. $\dfrac{C}{P}$
4. $\dfrac{A}{\pi l}$
5. $\dfrac{V^2}{2g}$
6. $\dfrac{I}{PT}$
7. $\dfrac{I}{P}$
8. $\dfrac{E}{I}$
9. DV
10. $\dfrac{PV}{R}$
11. $\dfrac{0\cdot 866}{d}$
12. $\dfrac{ts}{S}$
13. $\dfrac{4V}{\pi d^2}$
14. $\pi r^2 h$
15. $\dfrac{v-u}{a}$
16. $\dfrac{n-p}{c}$
17. $\dfrac{y-b}{a}$
18. $\dfrac{H-S}{L}$
19. $\dfrac{VR-2R}{V}$
20. $C(R+r)$
21. $\dfrac{S}{\pi r}-r$
22. $\dfrac{H}{ws}+t$
23. $2pC+n$
24. $\dfrac{D-TL}{12}$

Exercise 1.8

25. $\dfrac{R-2R}{V}$

26. $\dfrac{C-PC}{S}$

27. $\dfrac{v^2}{2g}$

28. $\sqrt{\dfrac{2E}{m}}$

29. $\sqrt{\dfrac{A}{\pi}}$

30. $\dfrac{w^2}{10^6}$

31. $\dfrac{t^2g}{4\pi^2}$

32. $\dfrac{4\pi^2W}{t^2g}$

33. $\sqrt{a^2-b^2}$

34. $\dfrac{D^2}{1\cdot44d}$

35. 125

36. 1·875

37. 240

38. 10

39. 30·18

40. 103·91

41. 3 375 000

42. 1982

Exercise 1.9

1. 9 mm

2. 1150 and 350 rev/min

3. 5·17 N

4. 63·1%

7. 196

9. 60

10. 2500

11. (a) 300 (b) £500 (c) £3000

Exercise 1.11

1. 12° 41′

2. 12·70 mm

3. 232·92 mm

4. 116·46

5. 138·56

Exercise 1.12

1. (a) 1·6794 (b) 1·9837 (c) 1·9958 (d) 1·9599 (e) 1·9927 (f) 1·0944
2. (a) 47° 40′ (b) 57° 19′ (c) 63° 49′
3. (a) 80·07 mm (b) 61·45 mm (c) 93·96 mm (d) 136·5 mm (e) 74·37 mm (f) 184·3 mm
4. (a) 30° 55′ (b) 23° 10′ (c) 63° 42′
5. 76·03 mm   6. 42° 54′   7. 166·8 mm   8. 77·50 mm   9. 217·4 mm

Exercise 1.13

1. (a) $\overline{1}$·6794 (b) $\overline{1}$·9837 (c) $\overline{1}$·9958 (d) $\overline{1}$·9599 (e) $\overline{1}$·9927 (f) $\overline{1}$·0944
   (g) 0·0239 (h) 0·8486 (i) 0·0087 (j) 0·0584 (k) 0·0743 (l) 0·4759
2. (a) 57° 2′ (b) 22° 25′ (c) 75° 7′ (d) 11° 42′ (e) 80° 2′ (f) 19° 50′
3. (a) 45° 43′ (b) 18° 35′ (c) 58° 1′
4. (a) 92·85 (b) 17° 38′

Exercise 1.14

1. 0, 100; 95·12, 30·90; 58·78, −80·90; −58·78, −80·90; −95·12, 30·90 mm
2. 14·29, 8·57; 35·72, 21·43 mm
3. 28·20, 10·26; 19·30, 22·97 mm
4. 160·7, 191·5; 191·5, −160·7; −160·7, −191·5; −191·5, 160·7 mm

Exercise 1.15

1. 5·32 mm

2. 1° 32′; 9·06; 11·74 mm   3. 44·30 mm

4. 5° 40′; 15·28 mm

5. 120·86 mm

6. 18·66 mm

Exercise 1.16

1. 94·19 mm

2. 45·28 mm

3. 43·58 mm

4. 11·19 mm

5. 77·27 mm

6. x = 17·5, y = 17·85, X = 47·5, Y = 14·06 mm

Exercise 1.17

1. 500·064 mm

2. 420·033 mm

3. 514·44 mm

4. 165·70 mm

Exercise 1.18

1. (a) C = 71°   b = 59·06 mm   c = 99·86 mm
   (b) A = 48°   a = 71·52 mm   c = 84·16 mm
   (c) B = 56°   a = 37·41 mm   b = 95·28 mm
   (d) A = 46°   b = 212·4 mm   c = 73·03 mm
   (e) C = 67°   a = 150·8 mm   c = 236·1 mm
   (f) C = 63° 32′   a = 94·00 mm   b = 114·6 mm
   (g) B = 135° 38′   a = 9·39 mm   c = 14·43 mm
   (h) B = 81° 54′   b = 99·47 mm   c = 36·09 mm
   (i) A = 53° 39′   a = 212·4 mm   b = 234·1 mm
   (j) B = 48° 31′   c = 26° 25′   c = 42·47 mm

2. 16·5 mm   3. 97·60 mm   4. 177·7 mm   5. 284·3, 138·7 mm
6. 173·2, 153·2, 187·9 mm.

Exercise 1.19

1. 653·5   2. 2·548   3. 3·738   4. 45·61
5. 75·45   6. (a) 0·4132 (b) 0·7939 (c) 32·16   7. 1·000
                     (d) 6·126   (e) 1·106   (f) 1·119
8. 25·399

Exercise 2.1

1. 7848 N   2. 183·5 kg   3. 573 K   4. 263 °C
5. 500 N/mm²   6. (a) 500 N/cm² (b) 0·05 N/mm² (c) 50 kN/m²
7. 3200 N   8. 2·44 N/mm²   9. 1·58 MN/m²   10. 94·26 kN
11. 56·56 MN   12. (a) 5656 N (b) 5027 N

Exercise 2.2

1. 300 J   2. 31·39 kJ   3. 588·6 W   4. 8·8 W
5. 833·3 W   6. 3·14 kW   7. 1·57 kW   8. 2·51 kW
9. 15 kW   10. 13·3 kW   11. 50 000 N/m²   12. 613 W
13. 0·31 W   14. 4·19 kW

Exercise 2.3

1. 1·837 mm   2. 24·998 mm   3. 0·078 mm   4. 93·4°C
5. 156·4°C   6. −70·9°C   7. 233·3 mm³
8. (a) 0·16 mm (b) 1508 mm³

Exercise 2.4

1. 3·312 MJ   2. 5·52 MJ   3. 1·46 MJ   4. 38·8°C
5. 17·7°C   6. 95·3 kg   7. 14·4 kW   8. 0·5 kg/s

Exercise 2.5

1. 25 560 kJ   2. 12 793 kJ   3. 9344 kJ   4. 12 065 kJ
5. 7511 kJ   6. 14 448 kJ

Exercise 2.6

1. 8·3°C   2. 5·62 bars   3. 1385 cm³   4. 66·7 kN/m²
5. 0·017 m³   6. (a) 5·6 m³ (b) 42 m³   7. 35·3 bar   8. 116 m³

Exercise 2.7
1. (a) 110 N at 43° to the horizontal.
   (c) 5385 N at 68° to the horizontal
2. 539 N at 68° to the horizontal
4. 20 kN
6. 1340 N at 63° to the horizontal
8. W = 330 N; R = 600 N
10. W = 870 N; R = 230 N
12. W = 1450 N; R = 680 N
14. 9330 N
16. 710 N

(b) 149 N at 16° to the horizontal.
(d) 211 N at 17° to the horizontal.
3. 5860 N and 4140 N
5. 230 N and 190 N
7. 11 700 N
9. W = 130 N; R = 240 N
11. 490 N
13. W = 3880 N; R = 1040 N
15. 250 N

Exercise 2.8
1. 112·5 N
5. 392 N; 589 N
9. 4000 N
13. 6000 N

2. 6000 N; 8000 N
6. 1580 N; 1120 N
10. 6667 N
14. 1800 N

3. 33 N; 333 N
7. 800 N; 2800 N
11. 160 N

4. R_1 = 750; R_2 = 2250
8. P = 300 N; Q = 500 N
12. 66·7 N

Exercise 3.2
1. 800 rev/min
2. 668 rev/min
3. 2000, 1540, 933, 720 rev/min
4. Lowest speed = 250 rev/min. Highest speed = 1000 rev/min
5. 140 rev/min
6. 26·8 rev/min
7. 982, 593, 1385, 1029, 1702, 2400, 933, 662, 400 rev/min
8. 25 mm
9. 3600 mm/min
10. 36 mm/min
11. (a) $\dfrac{50}{60}$ (b) $\dfrac{40}{30}$ (c) $\dfrac{50}{40}$ (d) $\dfrac{25 \times 50}{20 \times 30}$ (e) $\dfrac{50 \times 100}{30 \times 20}$

Exercise 3.3
1. 40·9 Nm
2. 785 W
3. 13·64 Nm; 136·4 Nm; 2728 N
4. 375 rev/min; 95·5 Nm
5. 15 mm/min; 1·21 MN
6. 528 kN

Exercise 3.4
5. 1·5; 1·25 min
6. 1·57, 480 mm
9. 5:1; 24 mm
10. 94·3 N
11. 55%

Exercise 4.1
1. 200 MN/m²
2. 200 MN/m²
3. 0·000 16
4. 0·000 33
5. (a) 592 MN/m² (b) 52%
6. (a) 250 MN/m² (b) 550 MN/m² (c) 26% (d) 74%
7. 0·5 mm
9. 200 000 MN/m²
10. 200 000 MN/m²
11. 12 670 N
12. (a) 64 MN/m² (b) 32 MN/m²
13. 37·7 kN
14. 38·9 kN
15. 0·8 MN/m²
16. 0·95 MN/m²

Index